KILAUEA VOLCANO, HAWAI'I

Kilauea is an active volcano in the Hawaiian Islands. It is one of five volcanoes that form the island of Hawai'i. Kilauea has been erupting continuously since 1983. It is an invaluable resource for volcanologists, who are scientists who study volcanoes.

The part of the volcano shown in the picture is called a spatter cone. It is made of spatter that forms when expanding gases in the lava separate the liquid rock into irregular shapes. The spatter falls back onto the ground. It builds up to form a spatter cone around a vent in a volcano.

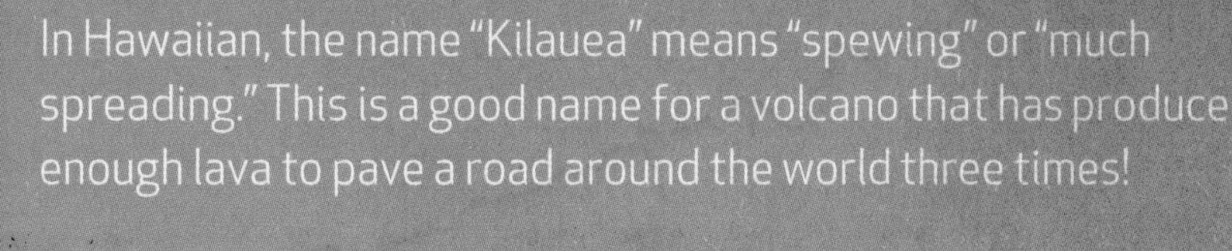

In Hawaiian, the name "Kilauea" means "spewing" or "much spreading." This is a good name for a volcano that has produced enough lava to pave a road around the world three times!

NATIONAL GEOGRAPHIC
SCIENCE

School Publishing

PROGRAM AUTHORS

Kathy Cabe Trundle, Ph.D.

Randy Bell, Ph.D.

Malcolm B. Butler, Ph.D.

Judith S. Lederman, Ph.D.

David W. Moore, Ph.D.

Program Authors

KATHY CABE TRUNDLE, PH.D.

Associate Professor of Early Childhood Science Education, The School of Teaching and Learning, The Ohio State University, Columbus, Ohio
SCIENCE

RANDY BELL, PH.D.

Associate Professor of Science Education, University of Virginia, Charlottesville, Virginia
SCIENCE

MALCOLM B. BUTLER, PH.D.

Associate Professor of Science Education, University of South Florida, St. Petersburg, Florida
SCIENCE

JUDITH SWEENEY LEDERMAN, PH.D.

Director of Teacher Education, Associate Professor of Science Education, Department of Mathematics and Science Education, Illinois Institute of Technology, Chicago, Illinois
SCIENCE

DAVID W. MOORE, PH.D.

Professor of Education, College of Teacher Education and Leadership, Arizona State University, Tempe, Arizona
LITERACY

Program Reviewers

Amani Abuhabsah
Teacher
Dawes Elementary
Chicago, IL

Maria Aida Alanis, Ph.D
Elementary Science
Intstructional Coordinator
Austin Independent
School District
Austin, TX

Jamillah Bakr
Science Mentor Teacher
Cambridge Public Schools
Cambridge, MA

Gwendolyn Battle-Lavert
Assistant Professor of Education
Indiana Wesleyan University
Marion, IN

Carmen Beadles
Retired Science Instructional
Coach
Dallas Independent
School District
Dallas, TX

Andrea Blake-Garrett, Ed.D.
Science Educational Consultant
Newark, NJ

Lori Bowen
Science Specialist
Fayette County Schools
Lexington, KY

Pamela Breitberg
Lead Science Teacher
Zapata Academy
Chicago, IL

Carol Brueggeman
K–5 Science/Math Resource
Teacher
District 11
Colorado Springs, CO

Program Reviewers continued
on page iv.

Acknowledgments

Grateful acknowledgment is given to the authors, artists, photographers, museums, publishers, and agents for permission to reprint copyrighted material. Every effort has been made to secure the appropriate permission. If any omissions have been made or if corrections are required, please contact the Publisher.

Illustrator Credits
All illustrations by Precision Graphics.
All maps by Mapping Specialists.

Photographic Credits
Front Cover G. Brad Lewis Photography.

Credits continue on page EM17.

National Geographic School Publishing
Hampton-Brown
www.myNGconnect.com

Printed in the USA.
RR Donnelley
Jefferson City, MO

ISBN: 978-0-7362-7717-4

11 12 13 14 15 16 17 18 19 20

5 6 7 8 9 10

Miranda Carpenter
Teacher, MS Academy Leader
Imagine School
Bradenton, FL

Samuel Carpenter
Teacher
Coonley Elementary
Chicago, IL

Diane E. Comstock
Science Resource Teacher
Cheyenne Mountain School District
Colorado Springs, CO

Kelly Culbert
K–5 Science Lab Teacher
Princeton Elementary
Orange County, FL

Karri Dawes
K–5 Science Instructional Support Teacher
Garland Independent School District
Garland, TX

Richard Day
Science Curriculum Specialist
Union Public Schools
Tulsa, OK

Michele DeMuro
Teacher/Educational Consultant
Monroe, NY

Richard Ellenburg
Science Lab Teacher
Camelot Elementary
Orlando, FL

Beth Faulkner
Brevard Public Schools
Elementary Training Cadre, Science Point of Contact, Teacher, NBCT
Apollo Elementary
Titusville, FL

Kim Feltre
Science Supervisor
Hillsborough School District
Newark, NJ

Judy Fisher
Elementary Curriculum Coordinator
Virginia Beach Schools
Virginia Beach, VA

Anne Z. Fleming
Teacher
Coonley Elementary
Chicago, IL

Becky Gill, Ed.D.
Principal/Elementary Science Coordinator
Hough Street Elementary
Barrington, IL

Rebecca Gorinac
Elementary Curriculum Director
Port Huron Area Schools
Port Huron, MI

Anne Grall Reichel Ed. D.
Educational Leadership/ Curriculum and Instruction Consultant
Barrington, IL

Mary Haskins, Ph.D.
Professor of Biology
Rockhurst University
Kansas City, MO

Arlene Hayman
Teacher
Paradise Public School District
Las Vegas, NV

DeLene Hoffner
Science Specialist, Science Methods Professor,
Regis University
Academy 20 School District
Colorado Springs, CO

Cindy Holman
District Science Resource Teacher
Jefferson County Public Schools
Louisville, KY

Sarah E. Jesse
Instructional Specialist for Hands-on Science
Rutherford County Schools
Murfreesboro, TN

Dianne Johnson
Science Curriculum Specialist
Buffalo City School District
Buffalo, NY

Kathleen Jordan
Teacher
Wolf Lake Elementary
Orlando, FL

Renee Kumiega
Teacher
Frontier Central School District
Hamburg, NY

Edel Maeder
K–12 Science Curriculum Coordinator
Greece Central School District
North Greece, NY

Trish Meegan
Lead Teacher
Coonley Elementary
Chicago, IL

Donna Melpolder
Science Resource Teacher
Chatham County Schools
Chatham, NC

Melissa Mishovsky
Science Lab Teacher
Palmetto Elementary
Orlando, FL

Nancy Moore
Educational Consultant
Port Stanley, Ontario, Canada

Melissa Ray
Teacher
Tyler Run Elementary
Powell, OH

Shelley Reinacher
Science Coach
Auburndale Central Elementary
Auburndale, FL

Kevin J. Richard
Science Education Consultant, Office of School Improvement
Michigan Department of Education
Lansing, MI

Cathe Ritz
Teacher
Louis Agassiz Elementary
Cleveland, OH

Rose Sedely
Science Teacher
Eustis Heights Elementary
Eustis, FL

Robert Sotak, Ed.D.
Science Program Director, Curriculum and Instruction
Everett Public Schools
Everett, WA

Karen Steele
Teacher
Salt Lake City School District
Salt Lake City, UT

Deborah S. Teuscher
Science Coach and Planetarium Director
Metropolitan School District of Pike Township
Indianapolis, IN

Michelle Thrift
Science Instructor
Durrance Elementary
Orlando, FL

Cathy Trent
Teacher
Ft. Myers Beach Elementary
Ft. Myers Beach, FL

Jennifer Turner
Teacher
PS 146
New York, NY

Flavia Valente
Teacher
Oak Hammock Elementary
Port St. Lucie, FL

Deborah Vannatter
District Coach, Science Specialist
Evansville Vanderburgh School Corporation
Evansville, IN

Katherine White
Science Coordinator
Milton Hershey School
Hershey, PA

Sandy Yellenberg
Science Coordinator
Santa Clara County Office of Education
Santa Clara, CA

Hillary Zeune de Soto
Science Strategist
Lunt Elementary
Las Vegas, NV

EARTH SCIENCE

CONTENTS

TECHTREK
myNGconnect.com

Student eEdition

Vocabulary Games

Digital Library

Enrichment Activities

CHAPTER 3

CHAPTER 4

CHAPTER 5

What Are Earth's Resources? 133

CHAPTER 6

How Do Weathering and Erosion Change the Land? 173

TECHTREK
myNGconnect.com

Student eEdition

Vocabulary Games

Digital Library

Enrichment Activities

CHAPTER 7

CHAPTER 8

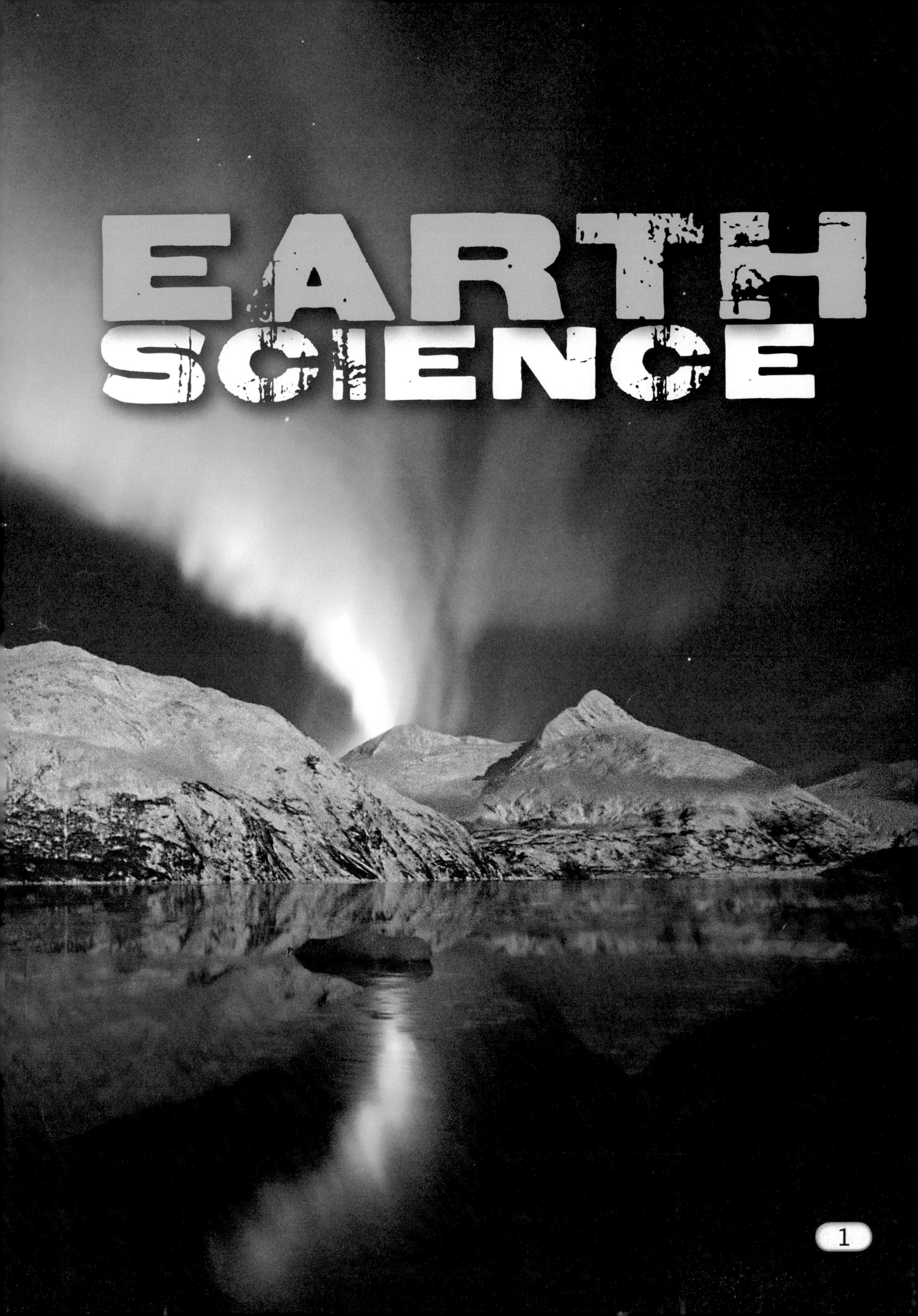

EARTH SCIENCE

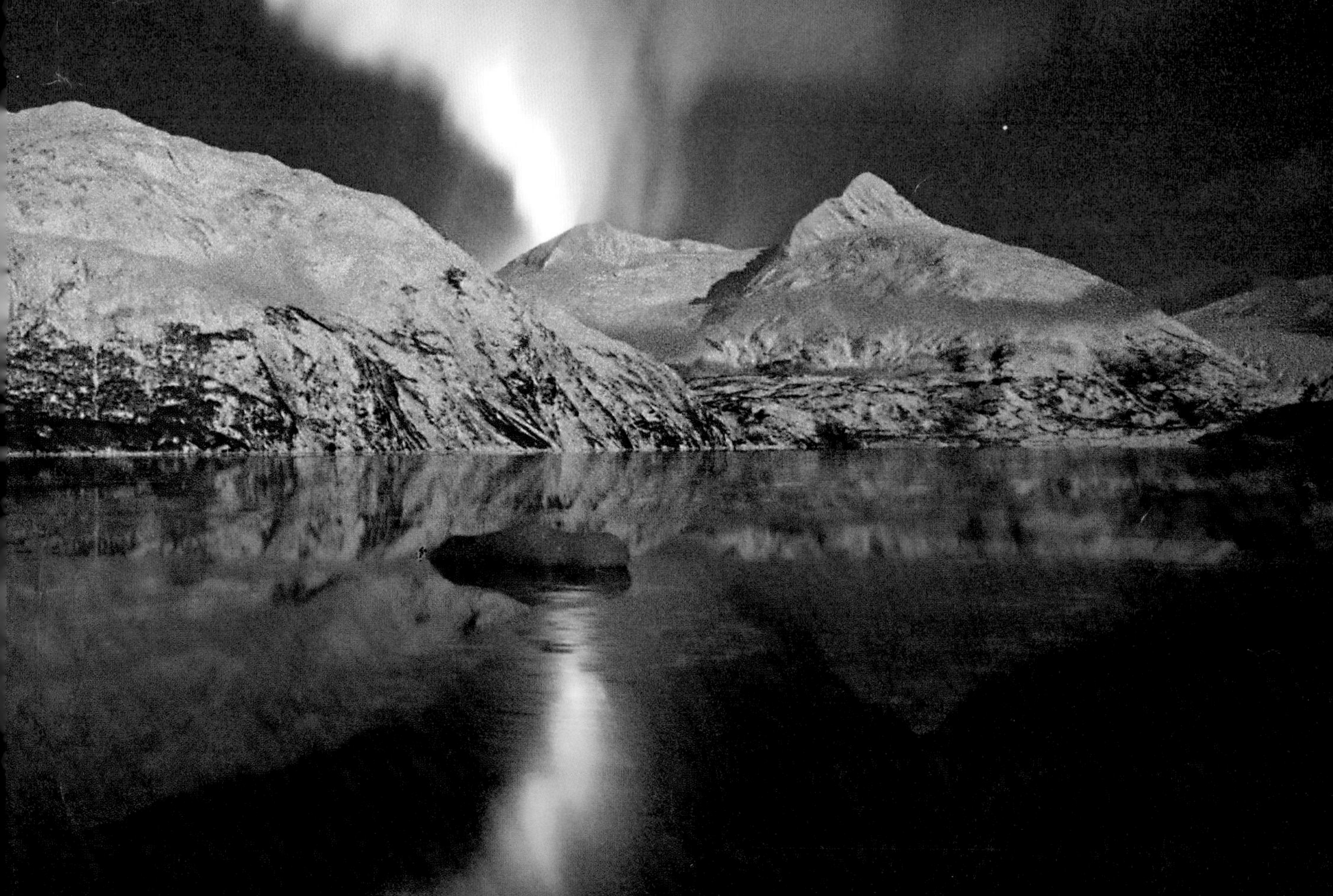

What Is Earth Science?

Earth science investigates all aspects of our home planet from its changing surface, to its rocks, minerals, water, and other resources. It also includes the study of Earth's atmosphere, weather and climates. As Earth is an object in space, Earth science also includes the study of Earth's relationship with the sun, moon, and stars. People who study our planet are called Earth scientists.

You will learn about these aspects of Earth science in this unit:

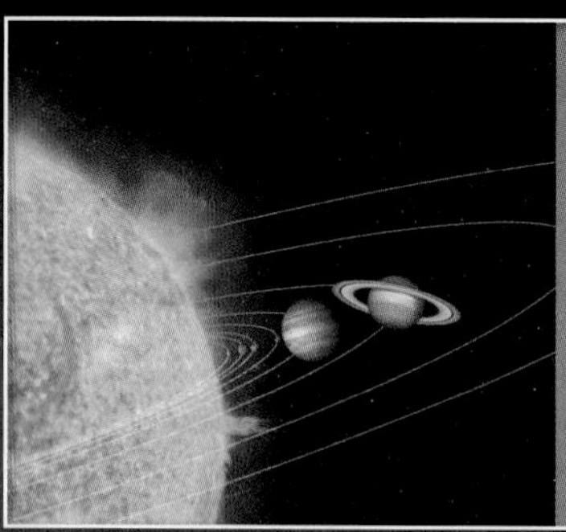

HOW ARE THE SUN, EARTH, AND MOON CONNECTED?

Earth revolves around the sun and rotates on its axis. The moon revolves around Earth. Earth scientists study the connections between the Earth, moon, and sun.

WHAT PROPERTIES CAN YOU OBSERVE ABOUT THE SUN?

The sun is a star. Its energy moves throughout the solar system causing objects on Earth to heat up. Its gravity works with that of Earth to keep Earth in orbit. Earth scientists study the sun to find out how it affects Earth.

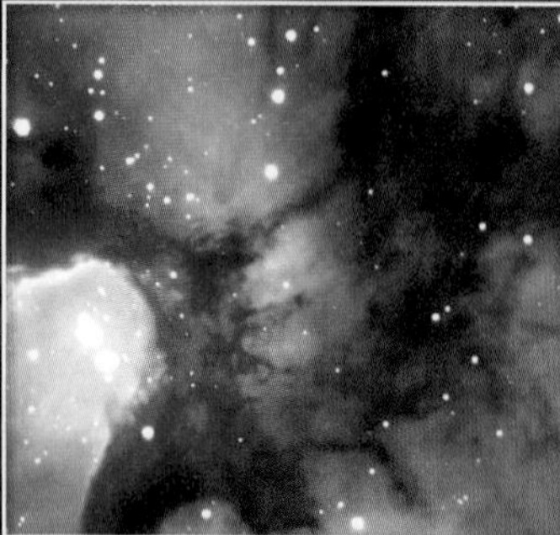

WHAT CAN YOU OBSERVE ABOUT STARS?

Earth scientists study the properties of stars, such as brightness, size, color, and temperature. These properties, along with how far away they are, can make stars look different from one another.

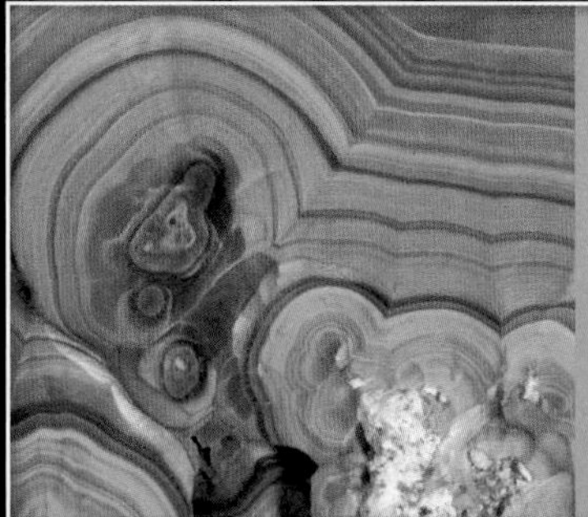

WHAT CAN YOU OBSERVE ABOUT EARTH'S MATERIALS?

Earth scientists study the materials that make up Earth, such as rocks and soils. Rocks and soils are made mostly of minerals. Earth scientists study the properties of minerals so they can classify them.

WHAT ARE EARTH'S RESOURCES?

Earth has all the resources people need to live. It has sunlight, plants, fossil fuels, and more. Some resources are renewable. Others are nonrenewable. Earth scientists study how people can use resources wisely.

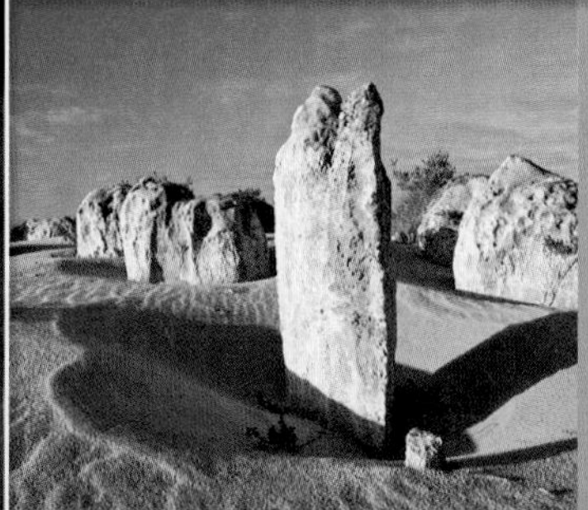

HOW DO WEATHERING AND EROSION CHANGE THE LAND?

Weathering breaks down or wears away rocks and other materials. Erosion carries these weathered materials from one place to another. Earth scientists study these processes and how they change the land.

HOW DOES EARTH'S SURFACE CHANGE QUICKLY?

Earthquakes, volcanoes, landslides, and weather events can change Earth's surface quickly. It is important for Earth scientists to study these processes so they can help people survive these natural events.

HOW DOES EARTH'S WATER MOVE AND CHANGE?

Water covers most of Earth's surface. As it moves through the water cycle, it can change to a solid, a liquid, or a gas. Earth scientists are finding ways for people to limit their use of water.

NATIONAL GEOGRAPHIC

MEET A SCIENTIST

Madhulika Guhathakurta: Astrophysicist

Madhulika Guhathakurta is an astrophysicist and the lead program scientist for NASA's Living With a Star (LWS) program. LWS focuses on understanding and ultimately predicting how the sun changes and what the effects of those changes are here on Earth. Simply put, Lika's research is about understanding "space weather" better.

Our sun is just one of billions of stars in the universe. However, the sun is important because it's the closest star to Earth. Without the sun, life on Earth would not exist.

Lika is also leading a worldwide effort known as the International Living With a Star (ILWS) program. This initiative is made up of all the space agencies of the world to contribute toward the scientific goal for space weather understanding.

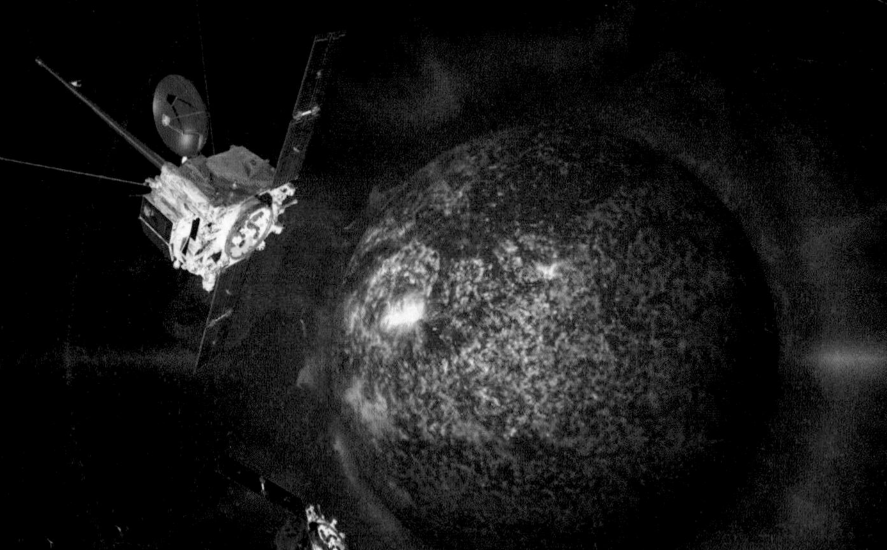

CHAPTER 1

HOW ARE THE SUN, EARTH, AND MOON CONNECTED?

On a usual day, the yellow sun and the pale moon are together in the sky. You can observe them both. The moon is always traveling around Earth. Earth is always spinning and traveling around the sun. These movements cause the many changes you see in the sky.

TECHTREK
myNGconnect.com

The sun lights up the sky during the day. The moon rises and glows, sometimes at night. Sometimes both the sun and moon are in the daytime sky.

6

7

After reading Chapter 1, you will be able to:

- Describe Earth as one of several planets that orbit the sun. **THE SUN AND THE PLANETS**
- Recognize that Earth rotates, causing day and night. **DAY AND NIGHT**
- Describe the apparent motion of the sun across the sky from east to west, generally during the day. **DAY AND NIGHT**
- Explain the changes in the length and direction of shadows throughout the day. **DAY AND NIGHT**
- Identify the moon as a satellite of Earth that reflects sunlight. **EARTH AND THE MOON**
- Recognize and describe the phases of the moon. **EARTH AND THE MOON**
- Describe the apparent motion of the moon across the sky. **EARTH AND THE MOON**
- Science in a Snap! Explain the changes in the length and direction of shadows throughout the day. **DAY AND NIGHT**

CHAPTER

1

HOW ARE THE SUN, E AND

On a usual day, the yellow sun and the pale moon are together in the sky. You can observe them both. The moon is always traveling around Earth. Earth is always spinning and traveling around the sun. These movements cause the many changes you see in the sky.

TECHTREK
myNGconnect.com

Student eEdition

Vocabulary Games

Digital Library

Enrichment Activities

ARTH, MOON CONNECTED?

The sun lights up the sky during the day. The moon rises and glows, sometimes at night. Sometimes both the sun and moon are in the daytime sky.

SCIENCE VOCABULARY

revolve (ri-VAWLV)

To **revolve** is to travel around another object in space. (p. 10)

Earth and other planets revolve around the sun.

orbit (OR-bit)

An **orbit** is the path Earth or another object takes in space as it revolves. (p. 10)

The planets orbit the sun.

rotate (RŌ-tāt)

To **rotate** is to spin around. (p. 12)

Day and night happen because Earth rotates.

orbit
(OR-bit)

phase
(FĀZ)

revolve
(ri-VAWLV)

rotate
(RŌ-tāt)

satellite
(SAT-il-līt)

satellite (SAT-il-īt)

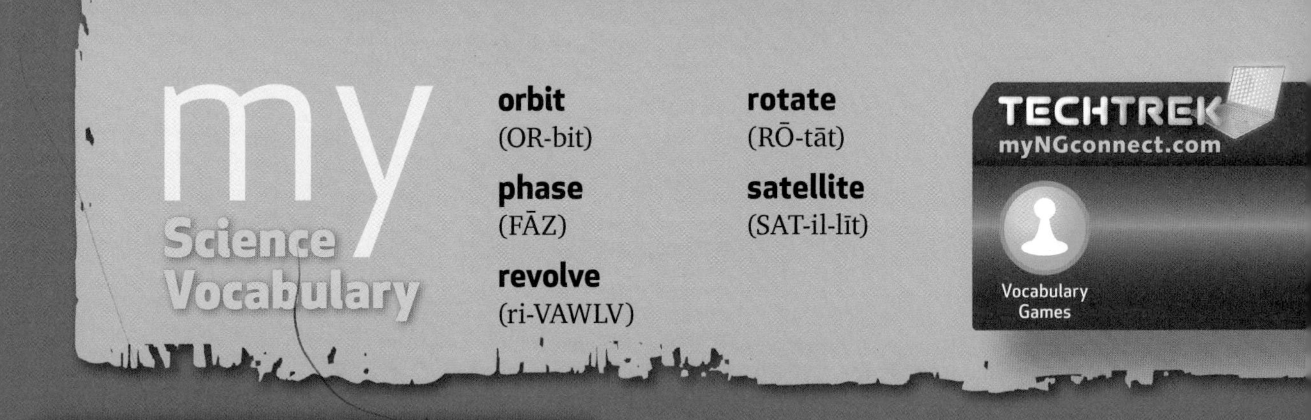

A **satellite** is an object that revolves around a larger object. (p. 16)

The moon is a satellite of Earth.

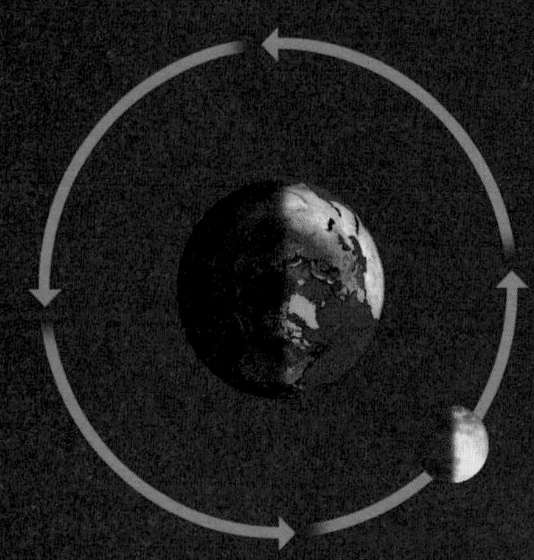

phase (FĀZ)

A **phase** is a lighted shape of the moon that we see from Earth. (p. 18)

A quarter moon is one phase of the moon.

The Sun and the Planets

Earth does not stay still. It **revolves**, or travels around, the sun. Earth moves very fast, but it takes one year to revolve once around the sun. In fact, a year is defined as the length of time it takes Earth to do this. The path Earth travels around the sun is called its **orbit**.

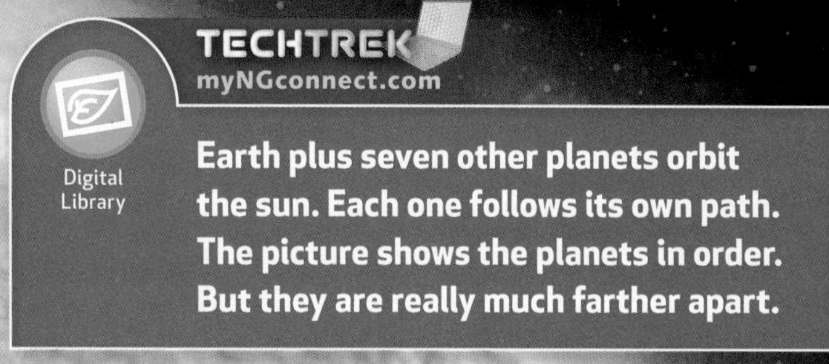

Earth plus seven other planets orbit the sun. Each one follows its own path. The picture shows the planets in order. But they are really much farther apart.

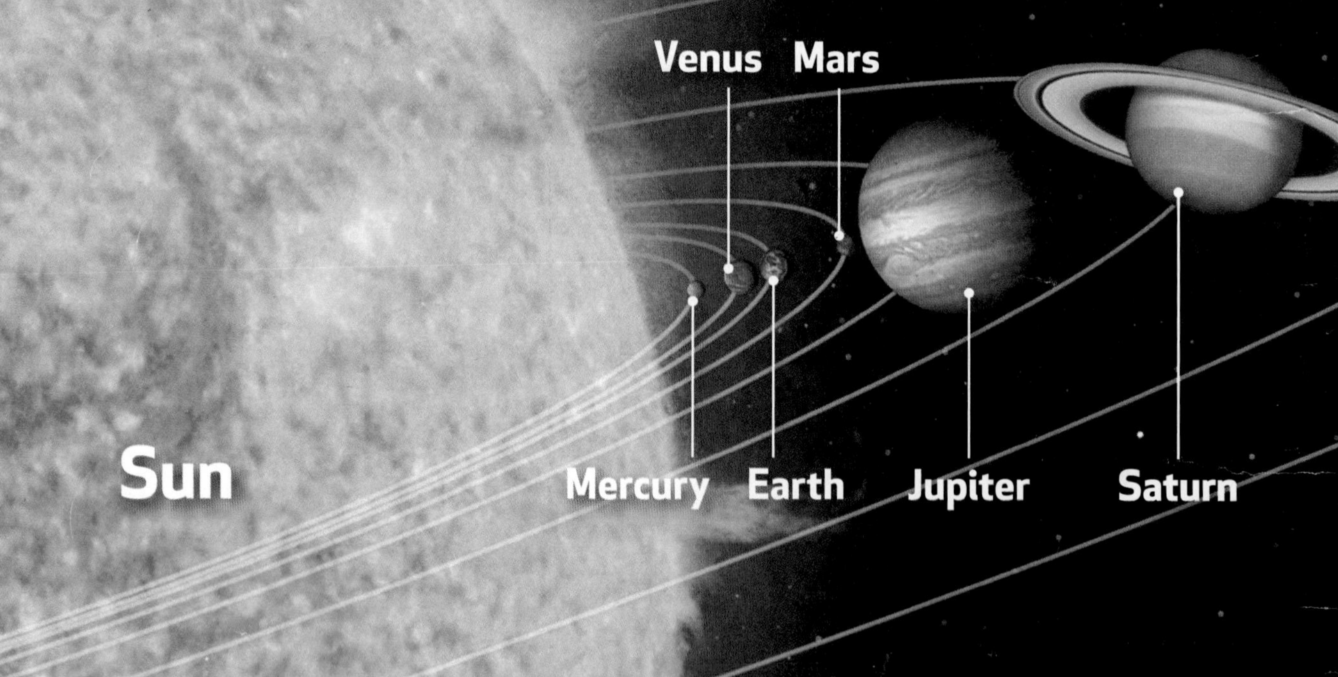

Earth is a planet. A planet is a huge sphere, or ball, that travels around a star. Earth is one of eight planets revolving around the sun. Mercury, Venus, and Mars are smaller than Earth. The other four planets are much bigger. Except for the sun, the eight planets are the largest objects in the solar system.

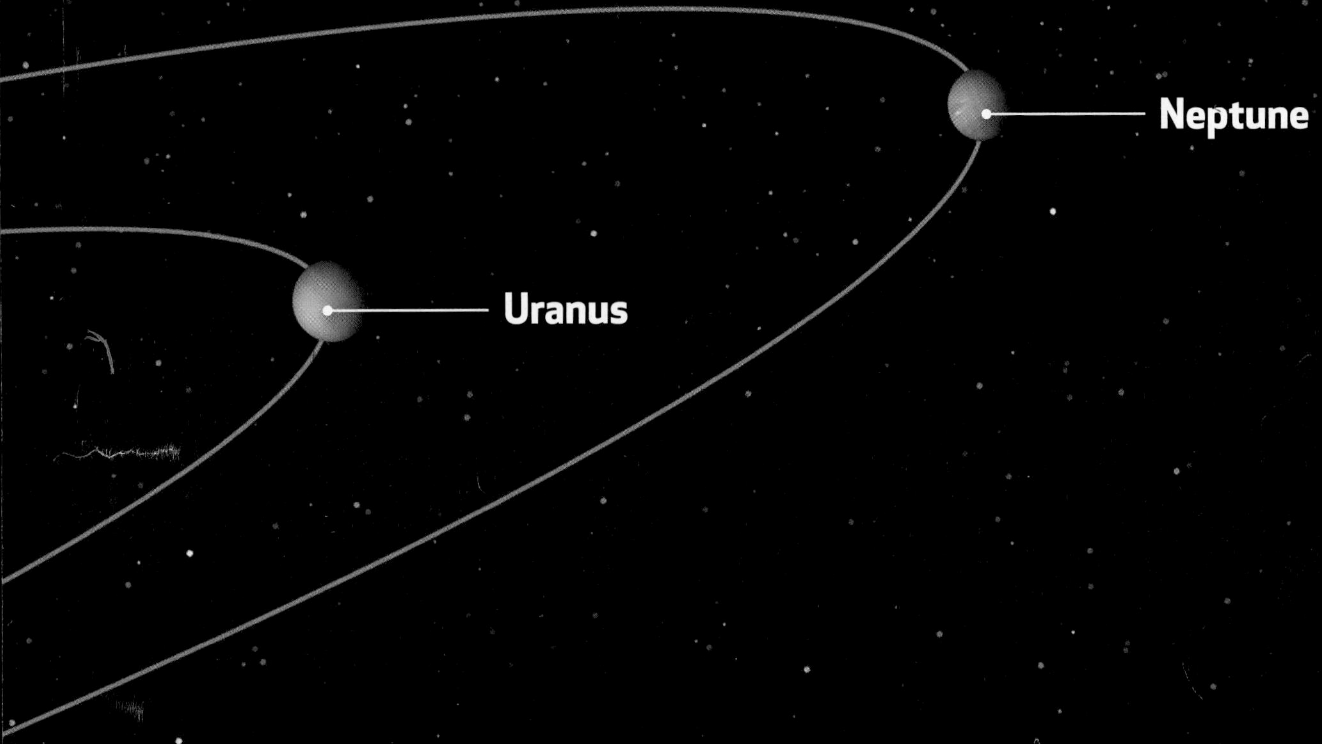

Neptune

Uranus

Before You Move On

1. What does *revolve* mean?
2. Look at the diagram of the solar system. How many planets are closer to the sun than Earth? How many are farther away?
3. **Draw Conclusions** Based on the diagram, does Neptune or Saturn have a longer orbit? Why do you think so?

Day and Night

As Earth revolves around the sun, it moves in another way, too. It **rotates**, or spins. Earth takes 24 hours to rotate once. Its rotation causes places on Earth to have day and night.

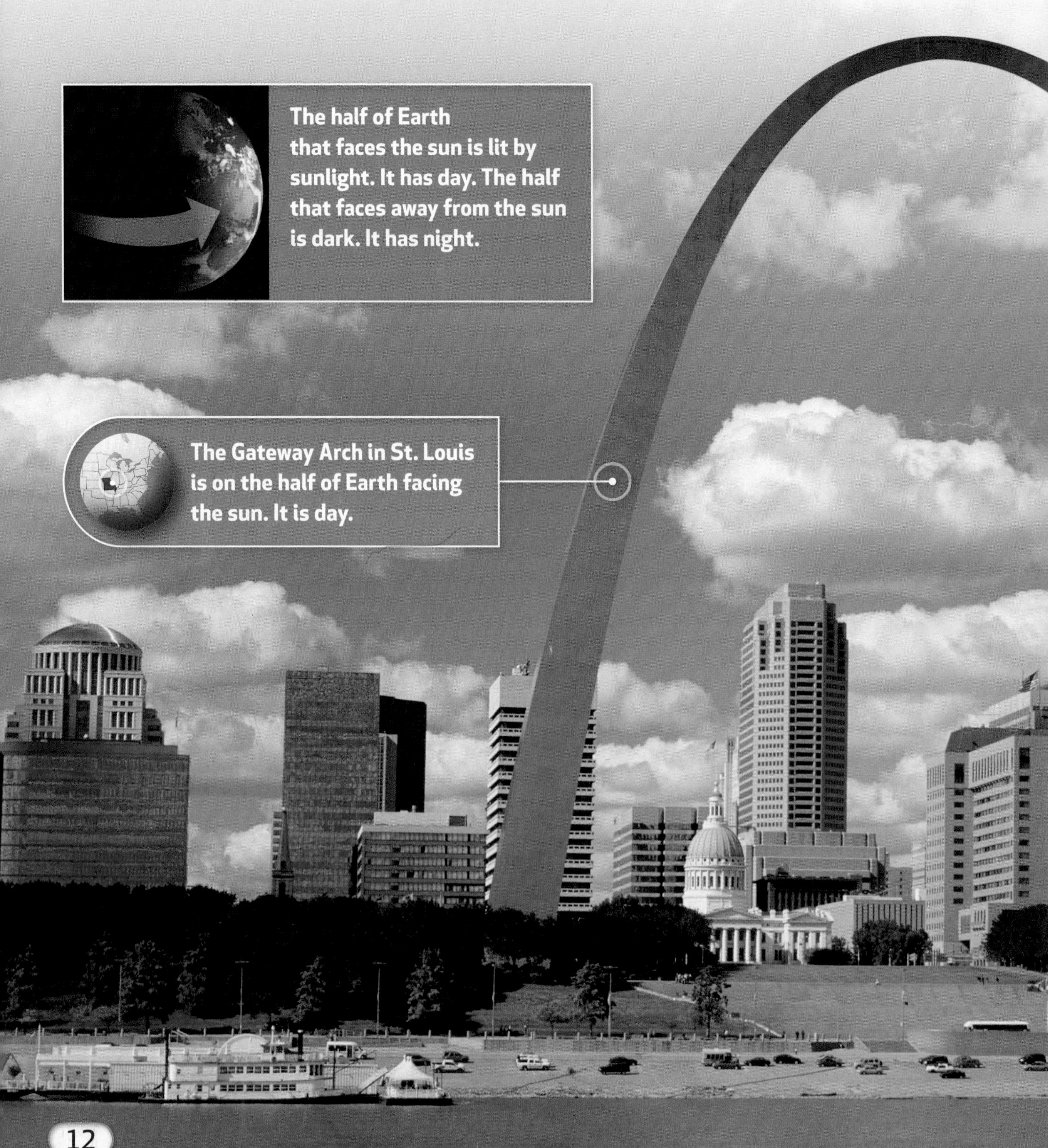

The half of Earth that faces the sun is lit by sunlight. It has day. The half that faces away from the sun is dark. It has night.

The Gateway Arch in St. Louis is on the half of Earth facing the sun. It is day.

Suppose you face a light in a dark room. Your front half is lit. Your back half is dark. What happens as you turn slowly around? Soon only half of your face is lit. Then none of your face is lit, and so on. The same thing happens to Earth. Half of Earth faces the sun at one time. As Earth spins, a different part of it faces the sun.

As Earth spins, the half of Earth that faced the sun now faces away. It has night. The half that faced away from the sun now faces toward it. It has day.

It is night. Which half of Earth is the Gateway Arch on now?

Earth never stops rotating. So places have a pattern of day and night, over and over. As St. Louis moves into the light, the sun seems to rise toward the east. During the day, it appears to move through the sky. As St. Louis moves into the darkness, the sun seems to set toward the west. But it's Earth moving past the sun, not the sun moving through the sky.

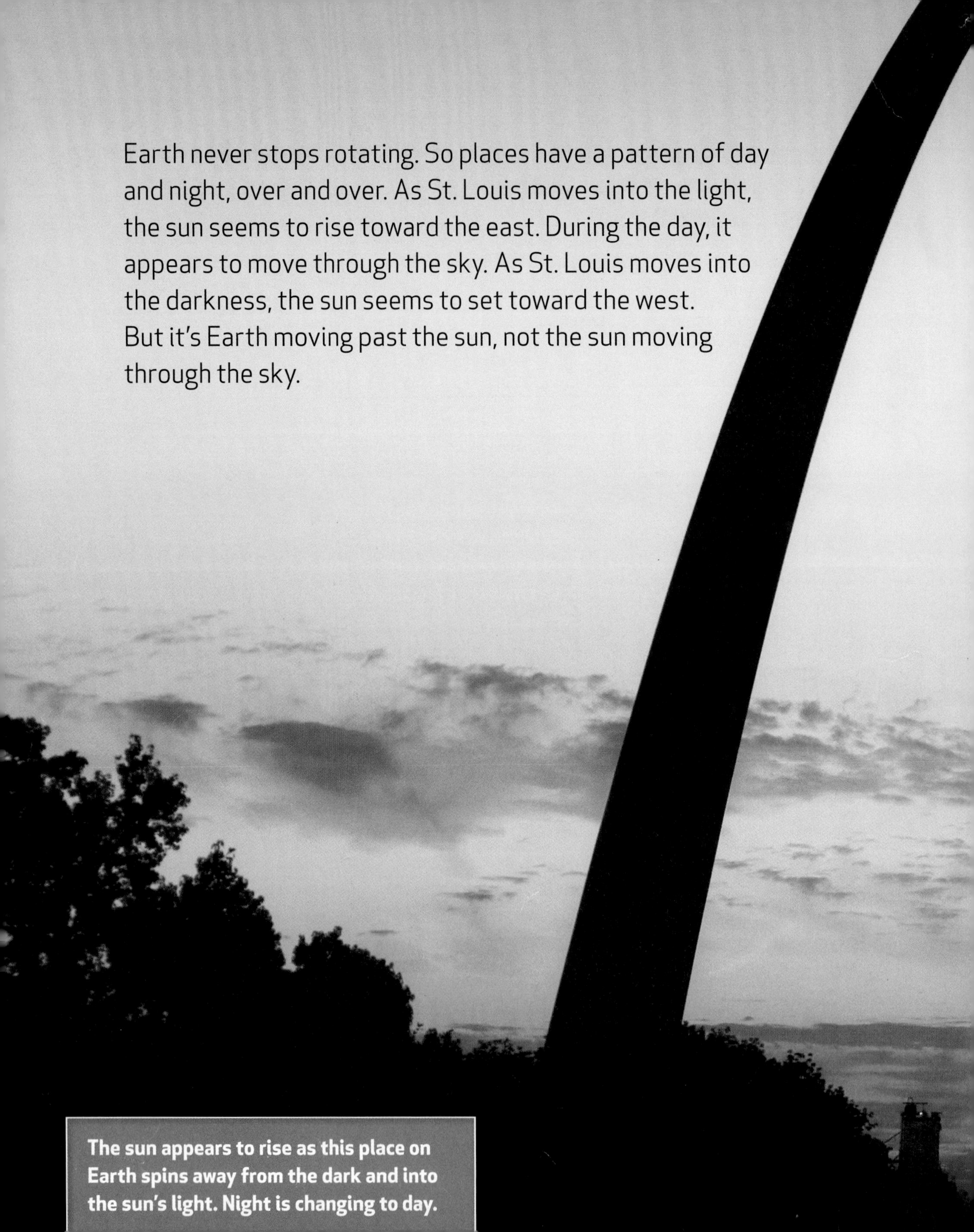

The sun appears to rise as this place on Earth spins away from the dark and into the sun's light. Night is changing to day.

In the early morning, the sun is low in the sky toward the east. Shadows are long and stretch toward the west. At midday the sun is high in the sky. Shadows are short. In the late afternoon, the sun is low in the sky toward the west. Shadows are long and stretch toward the east.

Science in a Snap! Moving Shadows

Make a model of the Gateway Arch.

Model how the sun appears to move. Observe your arch's shadow at morning, midday, and late afternoon.

How does the shadow change as the sun's position in the sky changes?

Before You Move On

1. What does *rotate* mean?
2. What pattern is caused by Earth's rotation?
3. **Draw Conclusions** How might people have used shadows before clocks were invented?

Earth and the Moon

The moon is Earth's closest neighbor. It is a sphere, or ball, that revolves around Earth. The moon is much smaller than Earth. A smaller object that orbits or moves around a larger object is called a **satellite**. The moon is a satellite of Earth.

The moon takes about one month to revolve completely around Earth. All the while, both Earth and the moon are rotating.

Like the sun, the moon seems to move through the sky. It looks like it is rising toward the east. It seems to set toward the west. Earth's rotation is the cause. Earth is spinning past the moon.

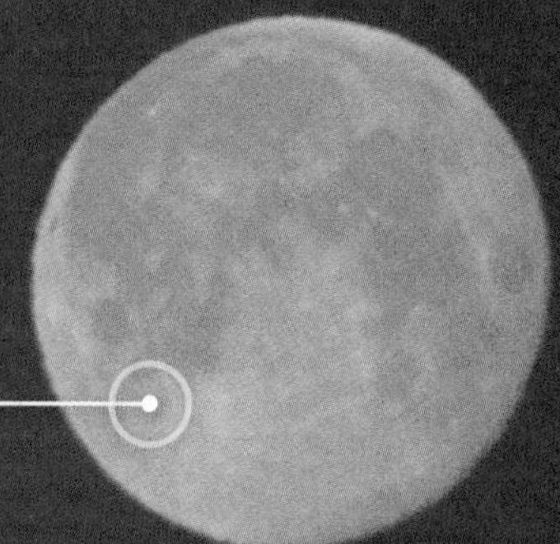

Night is not the only time to see the moon. You can also see it during the day.

Moonlight is really sunlight. The sun's light bounces off the moon and toward Earth.

Phases of the Moon Suppose you want to draw the moon. You might draw a circle. But you could draw a banana shape. During each month the moon's shape seems to change. Each lighted shape you see is a **phase** of the moon.

Different phases of the moon can often be seen during the day.

The sun lights up the half of the moon that faces it. As the moon revolves around Earth, you see different parts of the lighted half. Sometimes you see all of it. At other times you see only part of the lighted half. Sometimes you don't see any of it.

When the moon is a full circle, it is a full moon. If you are facing a full moon, where will the sun be?

The phases change in a pattern. You can see this pattern in the photos below. From one new moon to the next new moon takes about one month. You can see each phase at a certain time of the day or night.

TECHTREK
myNGconnect.com

Digital Library

PHASES OF THE MOON

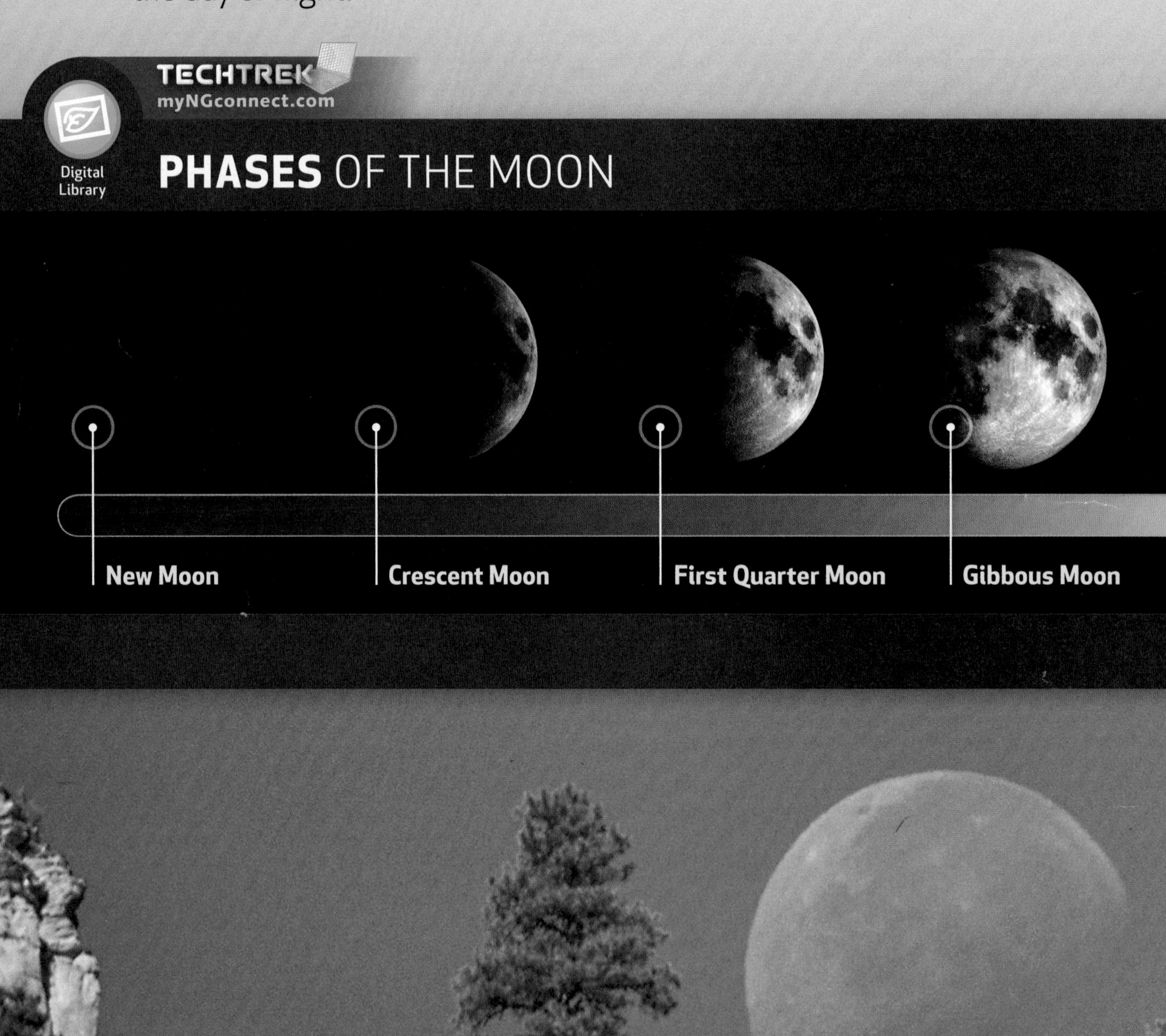

A new moon happens when the moon is lined up between the sun and Earth. In a new moon, the order is ***sun, moon, Earth.*** But when Earth is in line between the sun and the moon, then you see a full moon. The order is ***sun, Earth, moon.*** All other phases are stages leading up to a full moon or a new moon.

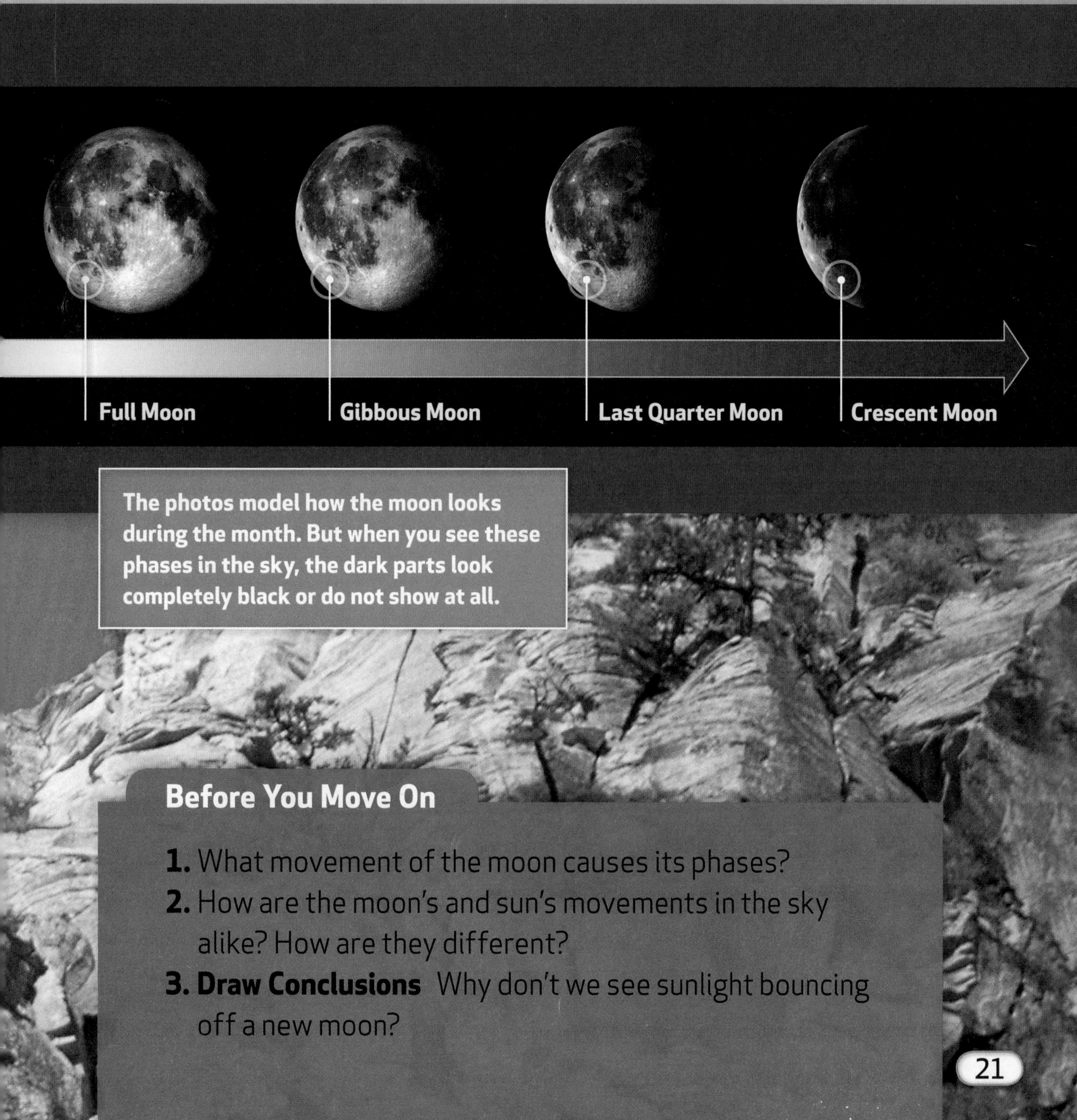

The photos model how the moon looks during the month. But when you see these phases in the sky, the dark parts look completely black or do not show at all.

Before You Move On

1. What movement of the moon causes its phases?
2. How are the moon's and sun's movements in the sky alike? How are they different?
3. **Draw Conclusions** Why don't we see sunlight bouncing off a new moon?

NATIONAL GEOGRAPHIC

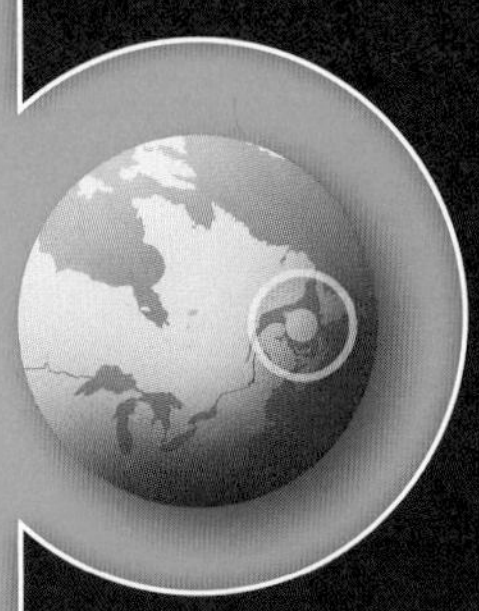

TIDES

THE MOON PULLS EARTH'S WATER

The daily rise and fall of ocean waters are known as tides. Along coasts, the water rises up over the shore and then slowly falls back again. When the water rises to its highest level, it is at high tide. When the water falls to its lowest level, it is at low tide.

Notice the dry land. It is low tide at Perce' Rock, Canada. The moon is pulling ocean water away from this spot.

What causes tides? The pull of the moon's and sun's gravity. High tide happens on the side of Earth that is facing the moon. The moon is pulling Earth's water. It is also high tide on the side farthest from the moon. The moon is pulling Earth away from the water on that side. Everywhere else has low tide.

As Earth rotates, the moon's gravity pulls in a different place. So a beach that had a high tide will have a low tide a few hours later.

About six hours later, the land is underwater. Earth has rotated one fourth of the way around. The moon is pulling the water toward this spot.

Conclusion

Earth is one of eight planets revolving around the sun. While Earth revolves, it also rotates. Its rotation causes day and night. It also causes the sun and moon to appear to move through the sky. The moon revolves around Earth like Earth revolves around the sun. As the moon revolves, you see different parts of its lighted side. These different parts you see are the phases of the moon.

Big Idea The movements of Earth and the moon around the sun cause regular changes you can see and predict.

Vocabulary Review

Match the following terms with the correct definition.

A. orbit
B. phase
C. revolve
D. rotate
E. satellite

1. To travel around another object in space
2. An object that revolves around a larger object
3. A lighted shape of the moon that we see from Earth
4. The path Earth or another object takes in space as it revolves
5. To spin around

Big Idea Review

1. **Recall** How many planets move around the sun?

2. **List** In what two ways does Earth move?

3. **Explain** Why is moonlight really sunlight?

4. **Cause and Effect** Explain why Earth's rotation causes day and night.

5. **Analyze** Where will Earth be in its orbit one year from now? Explain your answer.

6. **Draw Conclusions** Suppose your friend is standing outside. His shadow is long and stretches toward the west. About what time of day is it? Explain how you can tell.

Write About Moon Phases

Sequence Number any four moon phases in order. For each phase, write a statement that tells something about it.

NATIONAL GEOGRAPHIC

CHAPTER 1

EARTH SCIENCE EXPERT: ELECTRICAL ENGINEER

Engineers help scientists study Earth and space.

Engineers use scientific knowledge to solve practical problems. What is an electrical engineer? An electrical engineer can design new electronic technology that will solve problems.

Shannon Rodriguez-Sanabria is an electrical engineer at the National Aeronautics and Space Administration (NASA).

TECHTREK
myNGconnect.com

Digital Library

NASA uses Rodriguez-Sanabria's equipment to test ocean salt content, collect weather data, and run communications satellites.

TECHTREK
myNGconnect.com

Student eEdition

Digital Library

What do you need to study to become an electrical engineer?

In high school, take advanced math, chemistry, and physics classes. Always say "Yes!" to opportunities—for projects, fairs, research activities—that come your way. In college, study and get a degree in electrical engineering.

Do you see a strong connection between what you do and Earth science?

My job as an engineer enables scientists to do their jobs. Without the instruments that the engineers build there wouldn't be data for scientists to analyze. The instruments I work on help us study and understand the planet we live on.

What has been the coolest part of your career?

The coolest part of being an engineer at NASA is designing, developing, and testing new technology. I have also traveled to many cool places around the world, such as Costa Rica, Antarctica, and Argentina.

WHAT DO **ELECTRICAL ENGINEERS** DO?

1. First, electrical engineers decide how a piece of technology can solve a problem. They ask themselves, "What would this technology need to do?"
2. Next, they design the electrical circuits and other electronic parts for the technology. They often use computers to draw their designs.
3. Finally, the electrical engineers test their designs. They improve the technology if they can. If the technology doesn't work, the engineers try to figure out why and fix it.

NATIONAL GEOGRAPHIC

BECOME AN EXPERT

The Seasons: A Yearly Pattern

Day and night and how the moon changes during the month are patterns. Earth's **orbit** around the sun helps cause another pattern, the seasons. Read on to find out how the seasons change in places such as Shenandoah National Park in Virginia.

What season will you be in at this time next year? If you said the same one, you're right. The seasons make a pattern. The four seasons repeat year after year. The seasons are caused by the way Earth is tilted, or tipped to one side, as it **rotates**. Earth stays in this same tilted position throughout its orbit.

Summer in Virginia's Shenandoah National Park

orbit

An **orbit** is the path Earth or another object takes in space as it revolves.

rotates

To **rotate** is to spin around.

Summer In summer, Shenandoah National Park gets sunlight that is more straight on. So the park is heated by the sun most when it is summer. Study the diagram below. It shows you why this happens.

SUMMER

Summer is the warmest season. Why? Summer days are the longest of the year. The sun's energy is most straight on.

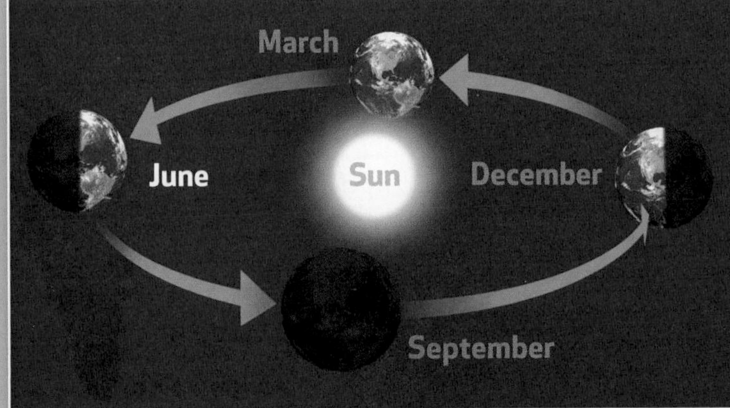

The sun's light is more straight on. So shadows are shortest in summer.

Fall Earth continues to **revolve** around the sun. Because of the way that Earth is tilted, the sun's energy hits Earth in a way that is less straight on. So the park is heated less. Temperatures become cooler in fall. Nights become longer.

FALL

Fall is cooler than summer. Why? Fall days are getting shorter. The sun's energy is less straight on.

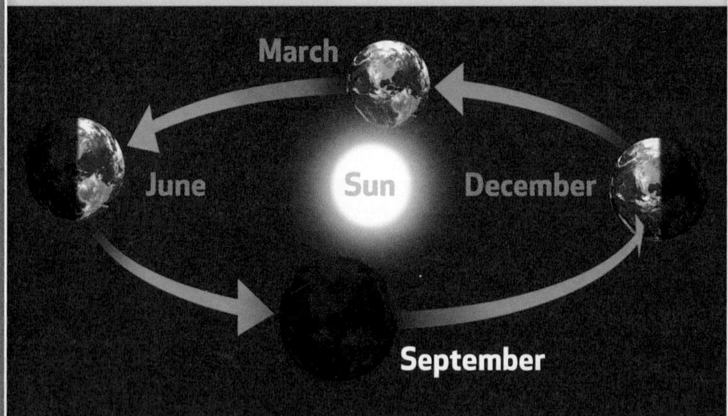

The sun's light is less straight on. Fall shadows are longer than summer shadows.

revolve

To **revolve** is to travel around another object in space.

Harvest Moon Earth's **satellite**, the moon, puts on a special show in the fall.

The first day of fall is around September 22. The full moon **phase** nearest that date is called a harvest moon. The harvest moon rises soon after sunset for several nights. It's so bright that farmers can still harvest crops after sunset.

Native Americans had different names for a harvest moon. Dying grass moon, travel moon, and blood moon are just a few.

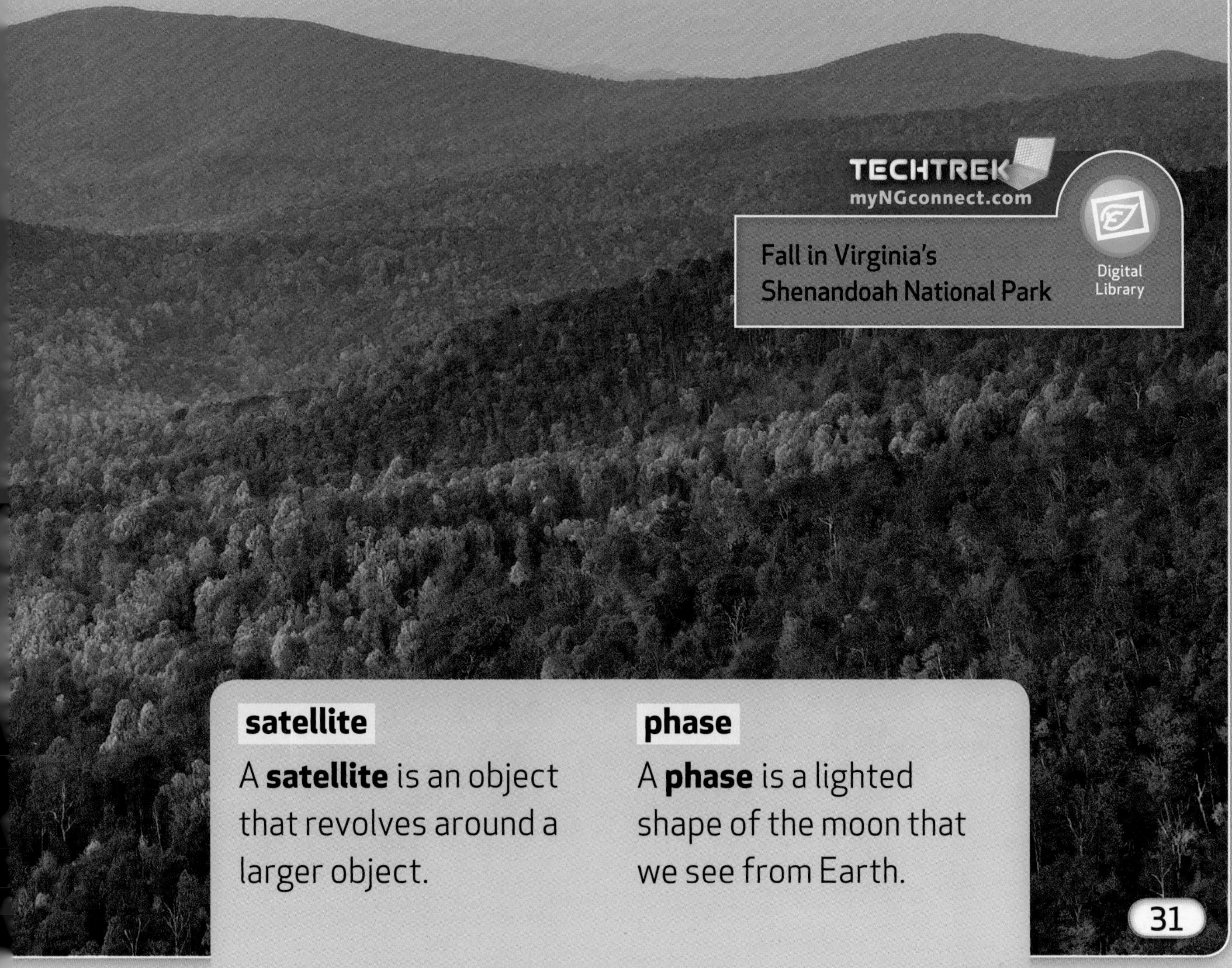

TECHTREK myNGconnect.com

Fall in Virginia's Shenandoah National Park

Digital Library

satellite

A **satellite** is an object that revolves around a larger object.

phase

A **phase** is a lighted shape of the moon that we see from Earth.

Winter In winter, Shenandoah National Park gets sunlight that is least straight on. So the park is heated by the sun least when it is winter. Temperatures are the coldest of the year. Days are the shortest.

Winter in Virginia's Shenandoah National Park

WINTER

Winter is coolest. Winter days are the shortest. The sun's energy is least straight on.

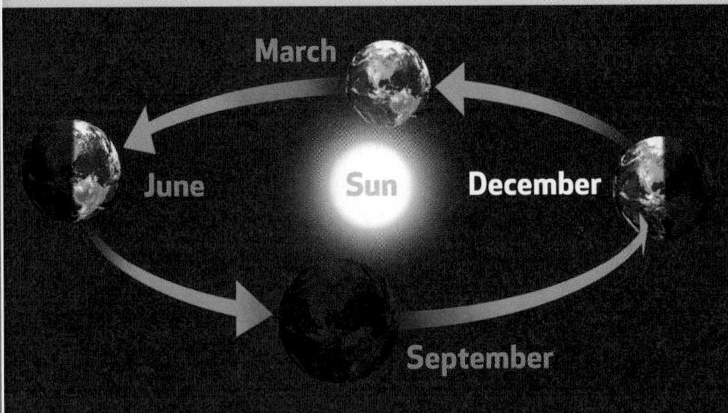

The sun's light is least straight on. So winter shadows are the longest.

Spring Earth continues revolving. Earth's tilt has not changed. But the sun's energy hits Earth in a way that is more straight on than in winter. So the park is heated more. Temperatures become warmer in spring and there is more daylight.

SPRING

Spring is warmer than winter. Why? Spring days are getting longer. The sun's energy is more straight on.

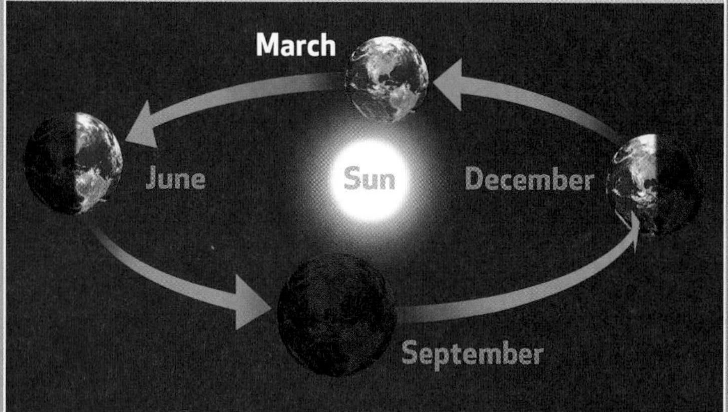

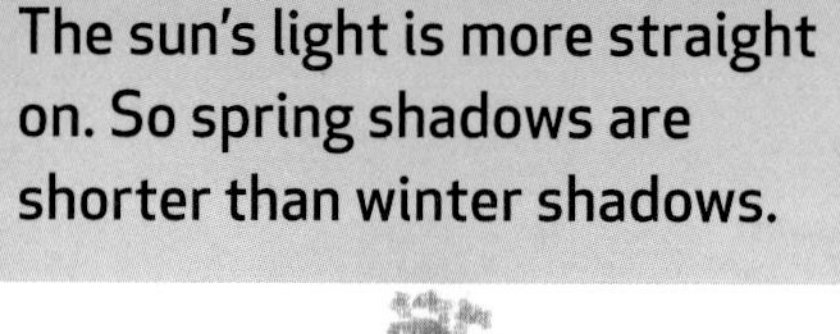

The sun's light is more straight on. So spring shadows are shorter than winter shadows.

The four seasons are a familiar pattern. They result from Earth's tilt in its orbit around the sun. You can see these changes year after year in Shenandoah National Park.

Spring in Virginia's Shenandoah National Park

CHAPTER 1 SHARE AND COMPARE

Turn and Talk What causes the changing seasons? Why are they a pattern? Form a complete answer to these questions together with a partner.

Read Select two pages in this section. Practice reading the pages. Then read them aloud to a partner. Talk about why the pages are interesting.

my SCIENCE notebook **Write** Write a conclusion that tells the important ideas about the pattern of seasons. State what you think is the Big Idea of this section. Share what you wrote with a classmate. Compare your conclusions.

my SCIENCE notebook **Draw** Form groups of three. Each person should choose one time of day—morning, noon, or late afternoon. Then draw a picture of yourself at that time. Write West on the left side of each paper and East on the right. Include the sun and your shadow in the drawing. Place the finished drawings in order of morning, noon, and late afternoon.

After reading Chapter 2, you will be able to:

- Describe the sun as a star that emits energy. **PROPERTIES OF THE SUN**
- Explain why the sun looks so large and bright from Earth. **PROPERTIES OF THE SUN**
- Recognize that the sun's energy can occur in the form of light. **PROPERTIES OF THE SUN**
- Demonstrate that radiant energy from the sun can heat objects. **THE SUN AS A SOURCE OF ENERGY**
- Identify the sun as the primary source of light and food energy on Earth. **PATHS OF THE SUN'S ENERGY**
- Science in a Snap! Demonstrate that radiant energy from the sun can heat objects. **THE SUN AS A SOURCE OF ENERGY**

CHAPTER

2

WHAT PROPERTIES CAN YOU OBSE ABOUT THE

High in the sky a bright, blazing star shines. You can see it in the morning and during the day. It is the sun. You can see the sun's light. Sunlight warms Earth's land, air, and water. You and almost all life on Earth depend on this one star.

TECHTREK
myNGconnect.com

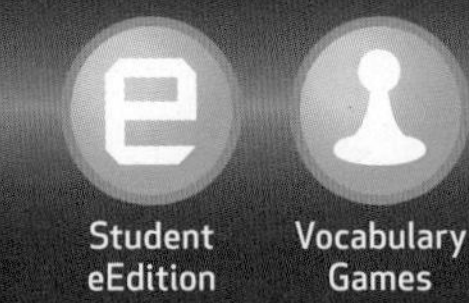

Student eEdition

Vocabulary Games

Digital Library

Enrichment Activities

RVE SUN?

The sun's light makes life on Earth possible.

SCIENCE VOCABULARY

sun (SUN)

The **sun** is the star that is nearest to Earth. (p. 42)

The sun looks bigger and brighter than other stars because it is the nearest star to Earth.

energy (EN-ur-jē)

Energy is the ability to do work or cause a change. (p. 44)

The sun sends energy out into space.

light (LĪT)

Light is a kind of energy you can see. (p. 45)

The Earth gets light energy from the sun.

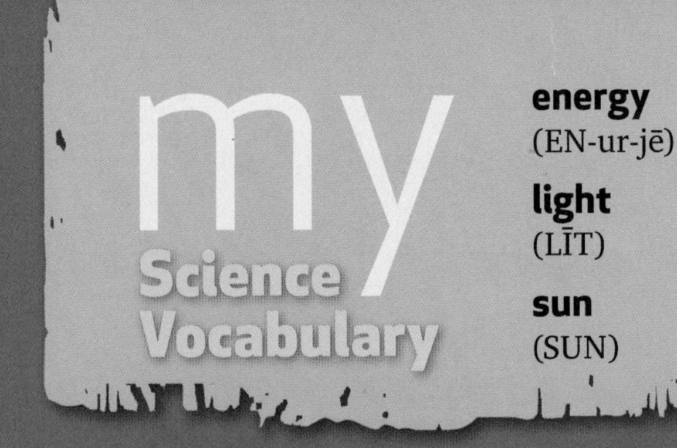

energy (EN-ur-jē)

light (LĪT)

sun (SUN)

temperature (TEM-pur-ah-chur)

transform (trans-FORM)

transform (trans-FORM)

To **transform** is to change. (p. 47)

Light energy can transform into heat energy and warm the land and other objects.

temperature (TEM-pur-ah-chur)

Temperature is a measure of how hot or cold something is. (p. 49)

The sidewalk has a cool temperature in the shade.

Properties of the Sun

The **sun** is just one of millions of stars in the sky. Why does it look bigger and brighter than other stars? The sun looks so large and bright because it is the nearest star to Earth. It is about 150 million kilometers (93 million miles) away. If you could drive to the sun in a car, it would take you about 177 years!

The Mars Rover took this photograph of the sun from the surface of Mars. The sun appears much smaller from Mars than it does from Earth. That's because Mars is farther away from the sun.

The sun is a medium-sized star. Some stars are much larger. Others are smaller. Compared to Earth, however, the sun is huge. More than 1 million Earths could fit inside it! As you can see in the model below, the sun is much bigger than even the largest planet, Jupiter.

TECHTREK myNGconnect.com

Digital Library

Look how small Earth is compared with the sun.

Life on Earth depends on energy from the sun. Energy is the ability to do work or cause a change. You need energy to live. The sun's energy is everywhere and changes form. The sun's energy is in the fuel used to run cars. It's in the food you eat. It's even in the air that moves around you as wind.

The sun is very bright. Do not look at it directly, or it will damage your eyes.

Different kinds of energy come from the sun. Some of the energy is in the form of **light**. Light is a kind of energy you can see. During the day, you can see sunlight all around you. When sunlight reaches Earth, it warms things. You can feel the sunlight warming your skin when you walk outside.

Before You Move On

1. Name a kind of energy that comes from the sun.
2. Why does the sun look bigger and brighter than other stars?
3. **Draw Conclusions** Why do you need the sun to live?

The Sun as a Source of Energy

Energy from the sun goes out into space. It moves out in all directions as light. Sunlight reaches throughout the solar system. The sun provides Earth and all the other planets with energy.

The sun provides energy to all the planets in the solar system, but not in equal amounts. Look at the diagram. Which planet do you think receives more of the sun's energy, Earth or Neptune? Why do you think so?

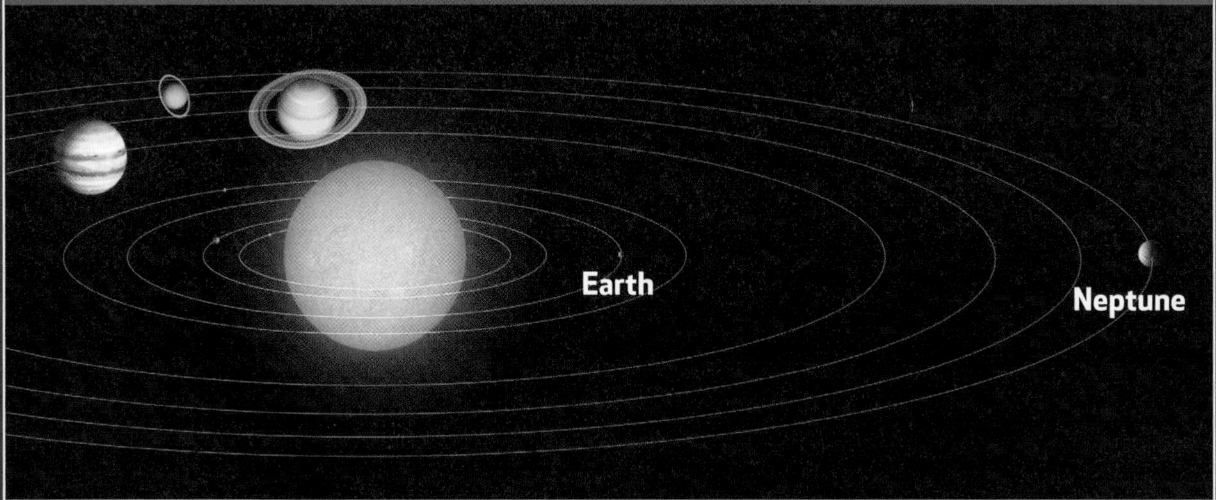

By providing light energy to Earth, the sun also provides heat. When light energy from the sun strikes Earth, it **transforms**, or changes, to heat energy. This heat warms the ground. It also then warms the air above the ground.

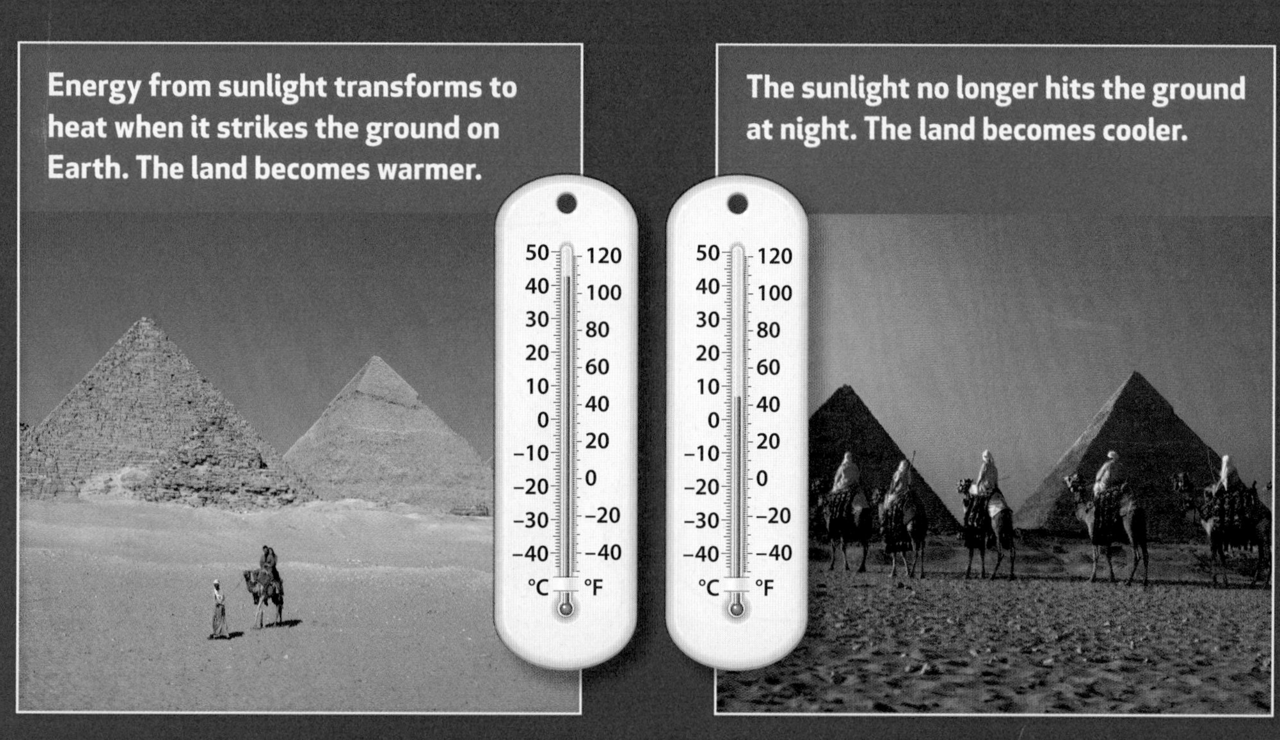

Energy from sunlight transforms to heat when it strikes the ground on Earth. The land becomes warmer.

The sunlight no longer hits the ground at night. The land becomes cooler.

Have you walked on a blacktop surface on a sunny summer afternoon? It's hot! Light energy from the sun has transformed to heat energy. It has warmed the blacktop.

Imagine walking on the same blacktop in the late evening. It's much cooler. The sun no longer shines on it. The blacktop has lost heat energy.

The long shadows show that it is late afternoon in this picture. Do you think that the blacktop road is hotter or cooler than it was at midday?

Trees provide shady spots on the black path. Is the surface hotter in the sun or in the shade?

The **temperature** of the blacktop changes during the day. Temperature is a measure of how hot or cold something is. When sunlight hits an object, heat energy causes that object's temperature to go up.

Science in a Snap! Compare Temperatures

Fold two sheets of black paper in half. Place one in the sun and one in the shade.

Wait 20 minutes. Place a thermometer inside each sheet of paper. Wait 1 minute. Record the temperatures.

What can you infer about how sunlight changes an object?

Before You Move On

1. What is temperature?
2. What effect does sunlight have on an object?
3. **Predict** The sun is shining on a piece of blacktop at 10 a.m. It is still shining on the same piece at 4 p.m. Will the blacktop be hotter in the morning or the afternoon? Why?

SUNSPOTS
STORMS ON THE SUN

From Earth, you see the outer layer of the sun. This layer is so bright that you cannot see past it to the inner layers. But not all of the sun's surface is bright. Dark blotches swirl and move across it. These sunspots are storms on the sun. Sunspots are darker than the rest of the surface because they are much cooler.

Even the smallest sunspots are more than 1,600 kilometers (1,000 miles) across. The largest are bigger than Earth.

Sunspots come and go. They grow and shrink. They appear alone or in large groups. Sunspots follow a pattern. The sun has a 22-year cycle, or pattern, of sunspots. During this cycle the number of sunspots increases and then decreases.

TECHTREK myNGconnect.com

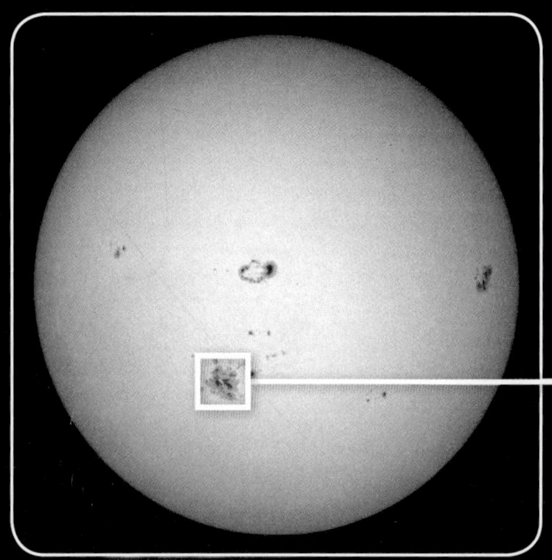

This picture shows a close-up of a sunspot.

Digital Library

The sunspot cycle begins with just a few sunspots. Over the next 11 years, their number grows. At the peak of the cycle, the sun may have as many as 100 sunspots. Through the next 11 years, many disappear. Then the cycle begins again.

Paths of the Sun's Energy

The sun's energy makes life on Earth possible. Living things need energy. Green plants use the energy of sunlight to make food. Animals eat plants to get this food energy, or they eat animals that have eaten plants. Without light from the sun, plants could not live. Without plants, animals could not live.

Light energy changes to food energy when an animal eats a plant.

The sun is the main source of light energy on Earth.

Sunlight also provides other energy that people use. Oil, natural gas, and coal are fossil fuels made from plants and animals that lived long ago. Solar panels change sunlight into electricity. Energy from the sun also creates wind. Windmills use wind to make electricity. People depend on light energy from the sun every day.

ENERGY FROM THE SUN

Plants change sunlight into food energy.

These panels take in sunlight. They help change it into electricity.

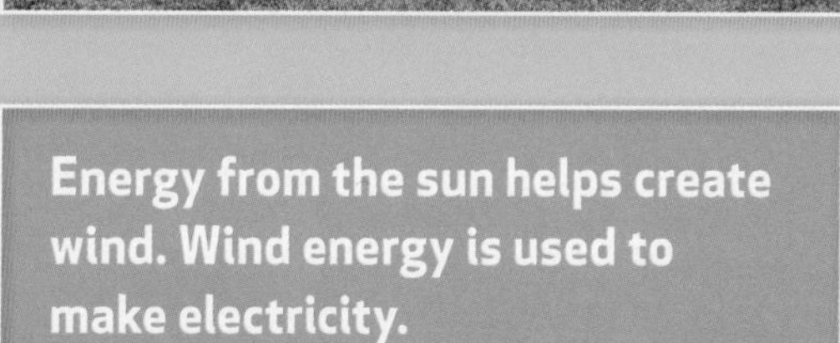

Light energy moved from sunlight to plants to animals to fossil fuels.

Energy from the sun helps create wind. Wind energy is used to make electricity.

Do you like corn on the cob? The energy you get from corn came from the sun. The energy used to bring you corn also came from the sun.

The corn plants change the sun's energy into food energy.

The truck will bring corn to the store. It runs on gas from oil, a fossil fuel. The fuel stored the sun's energy long ago.

Solar energy panels use sunlight to make electricity. The electricity lights the store.

The sun's energy takes many paths on Earth. It changes to different kinds of energy. People use these different energy sources for light, heat, food, and power for machines.

Natural gas is a fossil fuel. It heats and cooks the corn.

You eat the corn. It used to be the sun's energy. Now it's yours. Your body uses the energy to live, move, and grow.

Before You Move On

1. Where did fossil fuels come from?
2. What is the path that the sun's energy takes to a grazing cow?
3. **Generalize** What is the most important source of light and food energy on Earth?

Conclusion

The sun is the closest star to Earth. So the sun looks brighter and larger than other stars. The sun produces energy. Some of this energy is light that can transform to heat. Light energy from the sun is also transformed to food energy and other energy that people can use.

Big Idea The sun produces energy that provides light and heat to Earth.

The sun sends energy out into space and toward Earth.

Life on Earth uses light and heat energy from the sun.

Vocabulary Review

Match the following terms with the correct definition.

A. energy
B. light
C. sun
D. temperature
E. transform

1. A kind of energy you can see
2. A measure of how hot or cold something is
3. To change
4. The star that is nearest to Earth
5. The ability to do work or cause a change

Big Idea Review

1. **Define** What kind of energy does Earth get from the sun?

2. **Recall** The sun is just one star you see in the sky. Why does it appear so much bigger than the other stars?

3. **Cause and Effect** Suppose one bike is lying in the sun. Another bike is lying in the shade of a tree. Which bike will be cooler? Why?

4. **Explain** What happens when sunlight hits Earth?

5. **Generalize** Explain how the temperature of an object changes at different times during the day.

6. **Infer** Think of three ways you used energy from the sun today. Explain your answer.

Write About the Sun

Describe Millions of years from now, the sun will produce less energy. Write about what you think will happen to Earth.

CHAPTER 2

EARTH SCIENCE EXPERT: SOLAR ENGINEER

How can people use the sun's energy? Ask a solar engineer.

Much of the sun's energy bounces back into space. But what if people could collect more of the sun's energy? Then they could meet their energy needs without harming the environment. Solar engineer Chuck Kutscher is working to make this happen.

TECHTREK myNGconnect.com

Digital Library

Chuck Kutscher holds a piece of the shiny solar collector above him.

Kutscher leads a team of engineers. Together they use computers to help design new solar collectors. The solar collectors catch the sun's energy. They focus its energy on tubes, which become heated. People can use this heat energy to make electricity.

Kutscher loves solar energy research. Why? Solar energy, or energy from the sun, is friendly to the environment. It does not dirty the air, as burning coal and oil do. But that is not the only thing Kutscher likes about his work. Talking about solar energy to people is "the fun part of my job," he says.

To become a solar engineer, Kutscher studied math, science, and engineering. A student should also "be interested in exploring and learning new things," he says. "Wanting to help solve the world's energy problems is important, too."

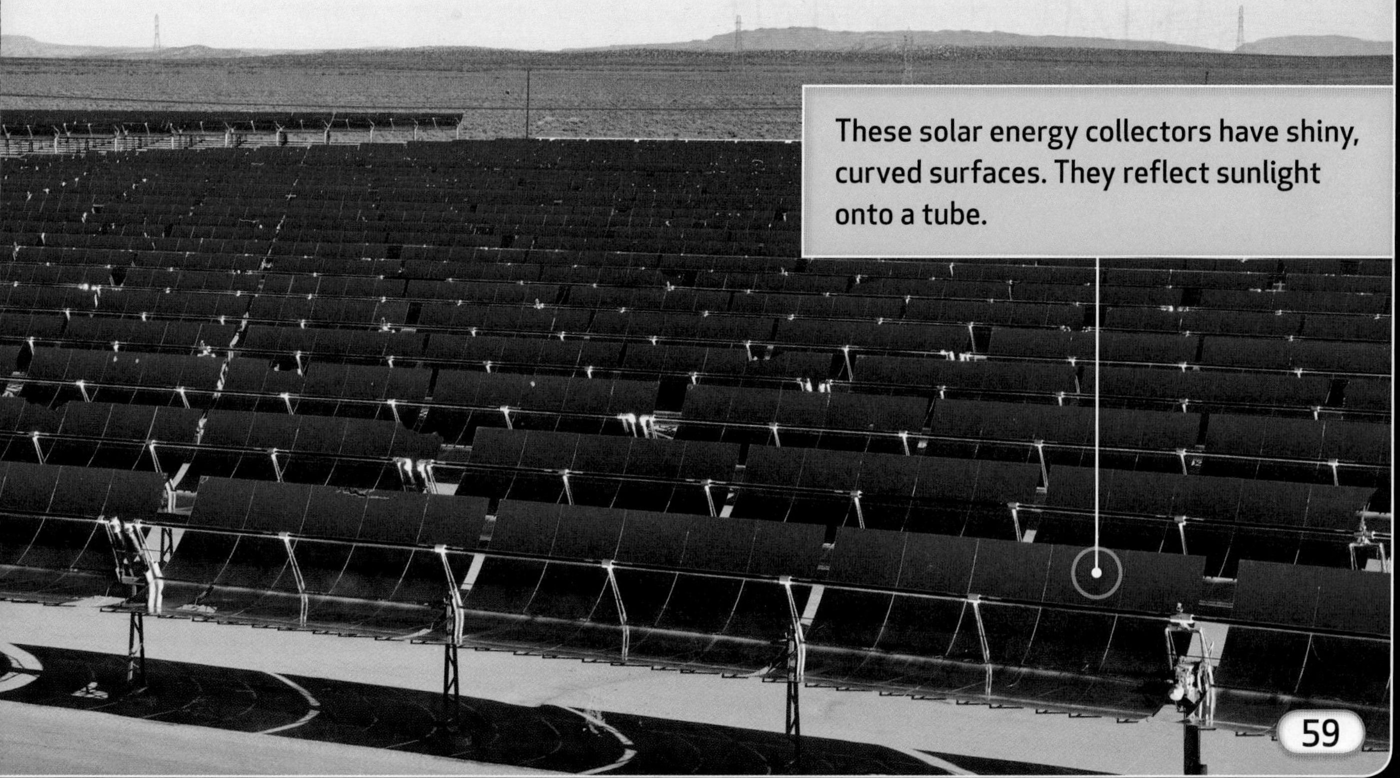

These solar energy collectors have shiny, curved surfaces. They reflect sunlight onto a tube.

NATIONAL GEOGRAPHIC

Solar Light Shows:
The Sun's Energy Colors the Sky

Earth depends on the **sun** for many things. **Energy** from the sun enables life to exist. But the sun can also provide colorful entertainment.

The sun is Earth's biggest source of energy.

sun

The **sun** is the star that is nearest to Earth.

Blue Sky Sunlight is white **light**. White light actually contains light of many different colors. Particles in Earth's air are just the right size to scatter the blue part of sunlight. When sunlight reaches the air, the particles bounce the blue light around. Blue light gets scattered throughout the sky. All the other colors continue on to Earth. That's why the sky looks blue to us.

WHY THE SKY IS BLUE

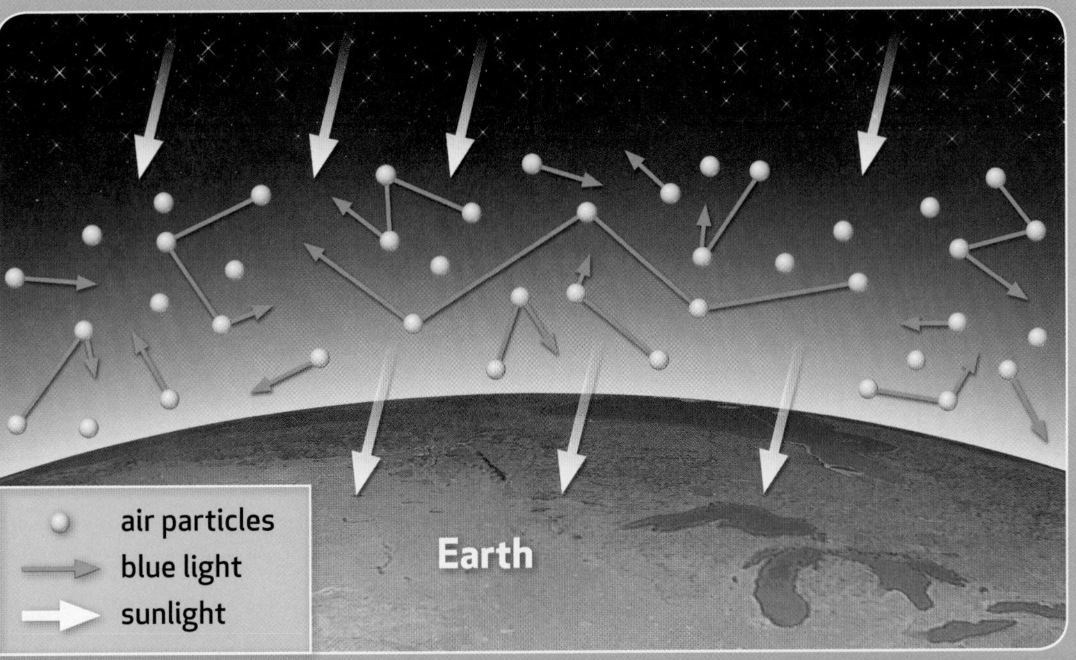

energy

Energy is the ability to do work or cause a change.

light

Light is a kind of energy you can see.

Rainbows Rainbows form when sunlight bends. Look at the picture of a prism. Remember, white light is made up of many colors. When white sunlight passes through the prism, the different colors separate into a rainbow.

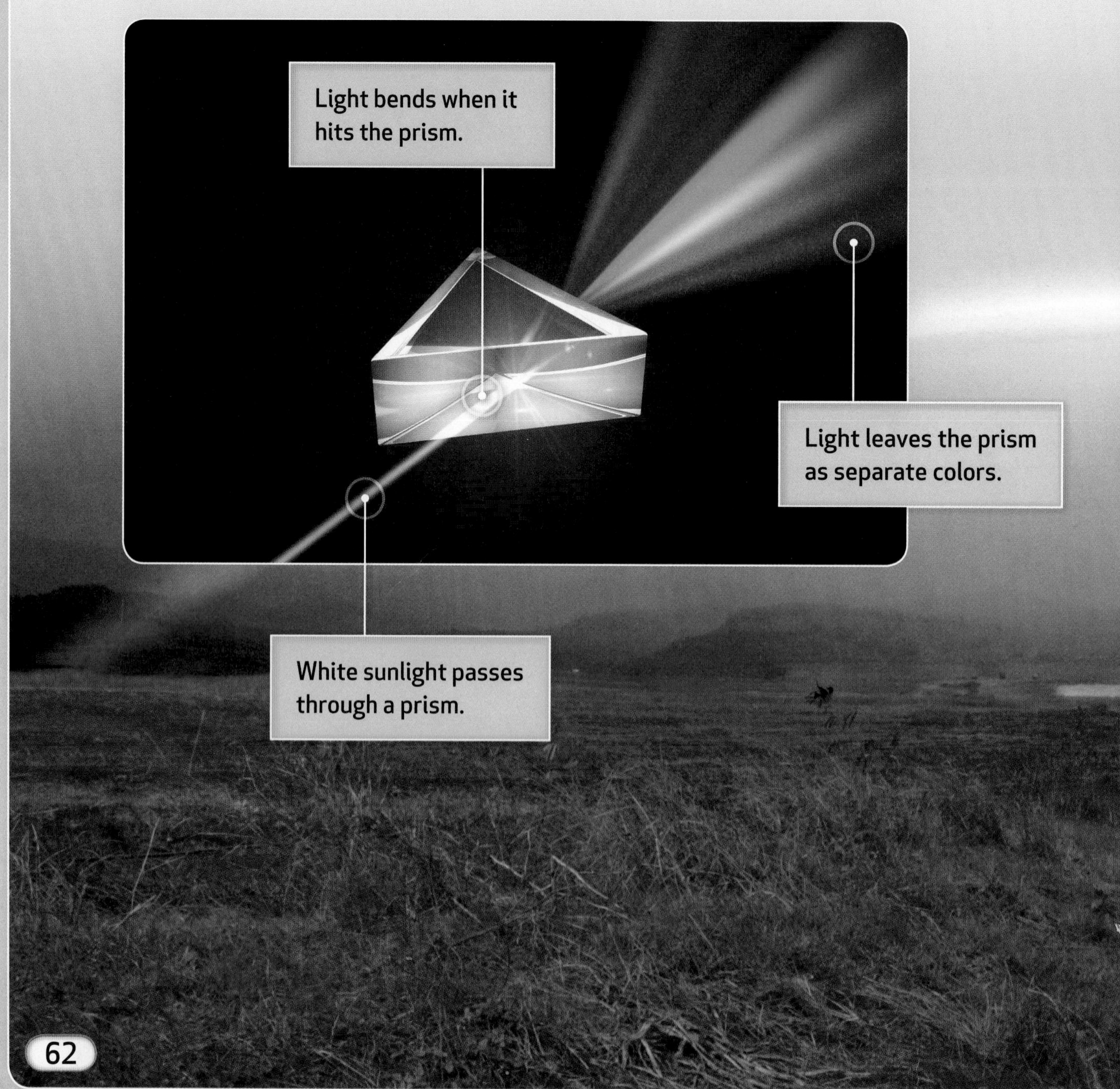

The same thing happens in the sky. Just after it rains, tiny drops of water are still in the air. The raindrops act like tiny prisms. White sunlight passes through raindrops. The different colors of light bend differently through the raindrops and separate. The colors form part of a circle through the sky, or a rainbow.

The colors in a rainbow always follow the same pattern: red, orange, yellow, green, blue, indigo, violet.

Halos When a storm is coming, sometimes high, wispy clouds move in first. The **temperature** of the air is so cold that the clouds are made of ice. These tiny ice pieces bend the sunlight. This bending forms a rainbow in the sky that looks like a circle around the sun.

Have you ever seen a ring of light around the sun? The ring is really in Earth's air, not around the sun.

temperature

Temperature is a measure of how hot or cold something is.

Sun Dogs Sometimes a full circle of light doesn't form, like it does with a halo. The ice crystals in the clouds bend sunlight in a different way. Then only small parts of a circle of light form. These parts form on either side of the sun and are called sun dogs. Sun dogs often happen when the sun is low in the sky.

Sometimes a sun dog appears on only one side of the sun. At other times a sun dog appears on both sides.

Auroras One of nature's most colorful light shows is called an aurora. Auroras often look like glowing curtains of light. Auroras happen when the sun's extreme heat produces tiny superhot particles. These particles shoot out into space. They crash into particles in Earth's air near the North and South poles. These crashes release energy that **transforms** into different colors of light.

TECHTREK
myNGconnect.com
Digital Library

Most auroras are green, red, purple, or white.

transform

To **transform** is to change.

The sun's radiant energy produces light shows on Earth. Halos and sun dogs appear when sunlight strikes tiny ice pieces in clouds. Rainbows happen when sunlight strikes raindrops. Auroras are caused by particles from the sun crashing into particles in Earth's air. A rare green flash appears when the sun sets in moist air.

This view of an aurora was taken by one of the crew on a space shuttle.

NATIONAL GEOGRAPHIC

CHAPTER 2 SHARE AND COMPARE

Turn and Talk What objects do you know that have been heated by the sun? Form a complete answer to this question together with a partner.

Read Select two pages in this section. Practice reading the pages. Then read them aloud to a partner. Talk about why the pages are interesting.

my SCIENCE notebook **Write** Write a conclusion that tells the important ideas about solar light shows. State what you think is the big idea of this section. Share what you wrote with a classmate. Compare your conclusions.

my SCIENCE notebook **Draw** Form groups of four. Have each person draw and color a picture of a solar light show. Label your drawing. Then write a caption explaining your drawing. Combine your drawings and make a poster. Don't forget to write a title for your poster. Share your work with other groups in your class.

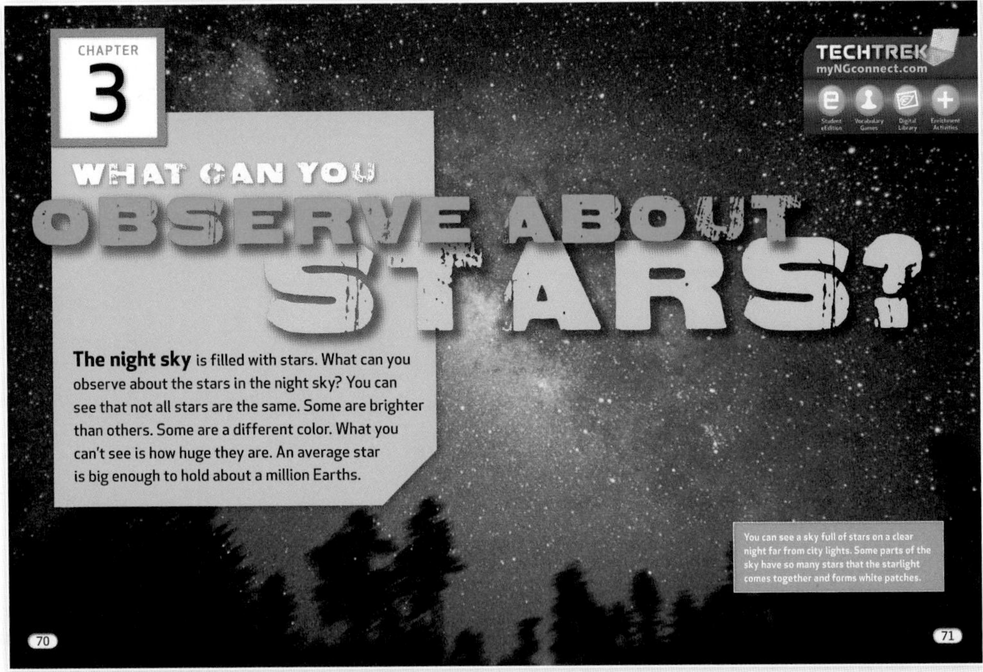
CHAPTER
3
WHAT CAN YOU
OBSERVE ABOUT
STARS?
The night sky is filled with stars. What can you observe about the stars in the night sky? You can see that not all stars are the same. Some are brighter than others. Some are a different color. What you can't see is how huge they are. An average star is big enough to hold about a million Earths.
TECHTREK
myNGconnect.com
You can see a sky full of stars on a clear night far from city lights. Some parts of the sky have so many stars that the starlight comes together and forms white patches.
70
71

After reading Chapter 3, you will be able to:

- Explain why stars are easier to see at night. **THE NIGHT SKY**
- Recognize that stars have different properties. **PROPERTIES OF STARS**
- Recognize that telescopes can be used to observe stars in more detail. **OBSERVING STARS**
- Describe how astronomers might use technology in their work. **OBSERVING STARS**
- Science in a Snap! Recognize that telescopes can be used to observe stars in more detail. **OBSERVING STARS**

CHAPTER

3

WHAT CAN YOU OBSERVE ST

The night sky is filled with stars. What can you observe about the stars in the night sky? You can see that not all stars are the same. Some are brighter than others. Some are a different color. What you can't see is how huge they are. An average star is big enough to hold about a million Earths.

TECHTREK
myNGconnect.com

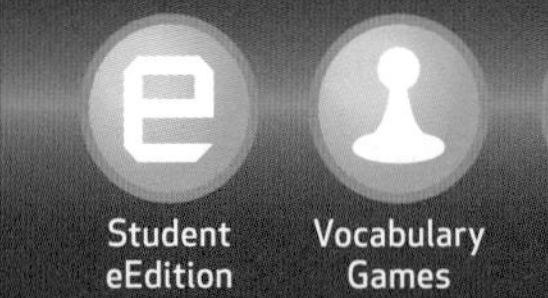

Student eEdition

Vocabulary Games

Digital Library

Enrichment Activities

ABOUT ARS?

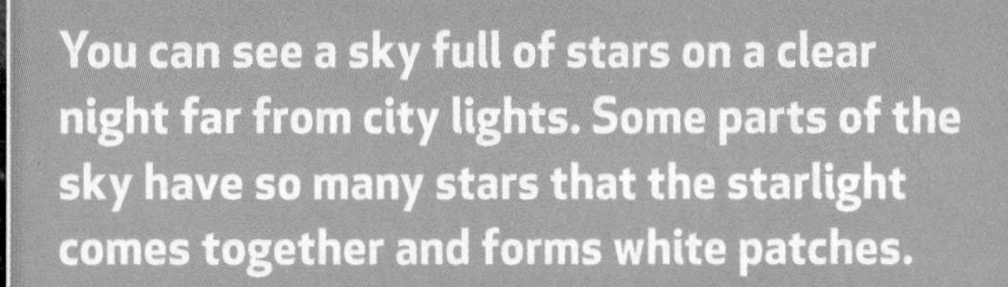

You can see a sky full of stars on a clear night far from city lights. Some parts of the sky have so many stars that the starlight comes together and forms white patches.

SCIENCE VOCABULARY

star (STAR)

A **star** is a glowing ball of hot gases. (p. 74)

Stars look like points of light because they are so far away from Earth.

property (PROP-ur-tē)

A **property** is something about an object that you can observe. (p. 76)

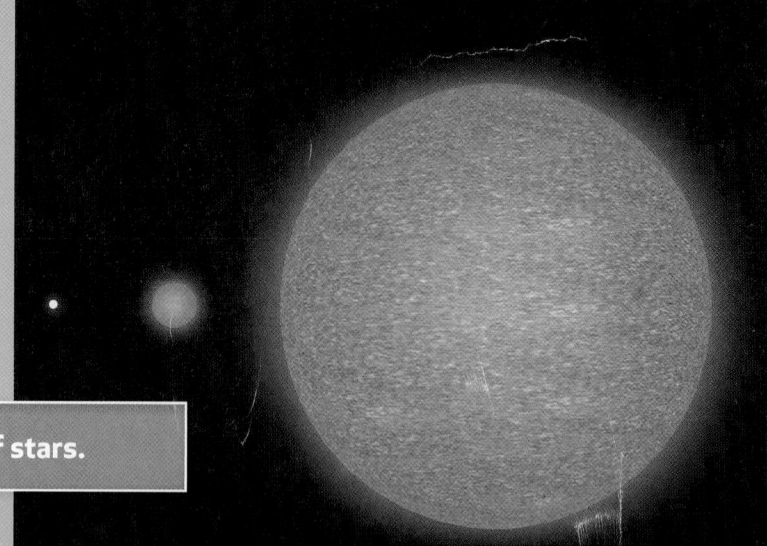

Size is a property of stars.

brightness (BRĪT-nes)

property (PROP-ur-tē)

star (STAR)

telescope (TEL-uh-scōp)

brightness (BRĪT-nes)

Brightness is the amount of light that reaches your eye from an object such as a star. (p. 76)

The brightness of a light depends partly on its distance from the viewer.

telescope (TEL-uh-scōp)

A **telescope** is a tool that magnifies objects and makes them look closer and bigger. (p. 82)

The telescopes of today are much more powerful than early telescopes.

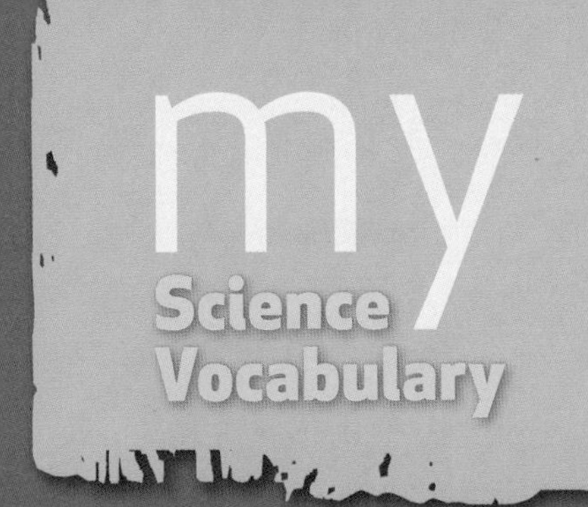

The Night Sky

Away from bright city lights, you can see many **stars** in the night sky. A star is a glowing ball of hot gases. Stars are so far away that they look like small points of light. In a city, you might see only the brightest stars at night. This is because streetlights and building lights wash out most of the starlight. So most stars are hard to see.

TECHTREK
myNGconnect.com

Digital Library

The sun sets on Easter Island in the Pacific Ocean. No stars can be seen yet because the sun's light is too bright.

The sun is a star. It is closer to Earth, so it looks much larger than other stars. You can see the sun only during the daytime. This is because during the day, your part of Earth faces the sun. At night, your part of Earth turns away from the sun. You can see the other stars in the sky then, because light from the sun no longer washes out the light from other stars.

As night begins to fall you can begin to see stars.

With no sunlight, the night sky is full of stars.

Before You Move On

1. What is a star?
2. Do you think it used to be easier for people to see the stars at night? Why or why not?
3. **Infer** The planet Jupiter is farther away from the sun than Earth. What do you think the sun would look like from Jupiter?

Properties of Stars

Brightness and Size Scientists classify, or group, stars by their **properties**. A property is something about an object that you can observe with your senses. Two properties of stars are **brightness** and size. Brightness is the amount of light that reaches your eye from an object such as a star.

Some stars look brighter than others because they are closer to Earth. It's like a line of streetlights. The lights are all the same size and give off the same amount of light. But the closer ones look brighter.

Why do some of these streetlights look brighter than others?

Unlike the streetlights, all stars are not the same size. Some look brighter because they are larger. In fact, some stars are called giants or supergiants. Others are much smaller and are called dwarf stars.

Some small stars are about the size of Earth. Some large stars are much larger than the sun. The sun is a medium-size star. This model compares it to some larger stars.

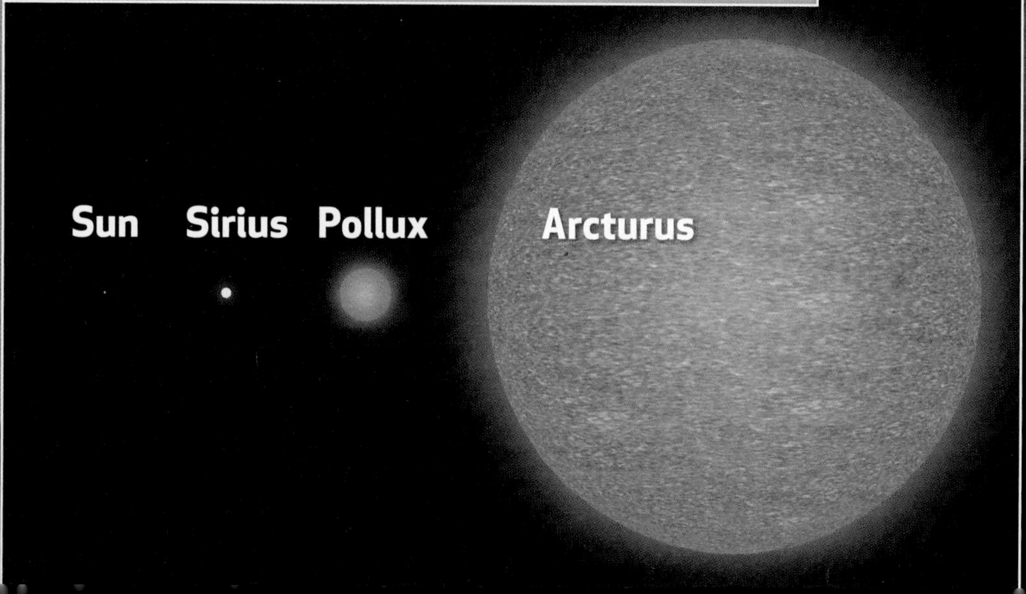

Temperature Size plays a part in how bright a star is. But the temperature of the star is important too. Temperature is how hot or cold something is. All stars are hot, but some are hotter than others. How can you tell which stars are hotter? It's all about color. Look at the metal pot in the picture. The pot is black. As it heats up, it changes color.

As the bottom of the pot heats up, it changes color. The temperature shows you how hot the pot is. If it could keep getting hotter, the whole pot would glow blue.

At first, the black pot glows red. As it gets hotter and hotter, the bottom of the pot changes color. It goes from red to orange to yellow to white to blue. Objects that glow red are cooler than those that glow white or blue.

Color At first glance, you may think all stars are white or light yellow. But look closely. Stars are different colors. And like the metal pot, a star's color is a clue to how hot it is. The coolest stars are red. The hottest stars are blue. White or blue stars are brighter because they are hotter and give off the most energy for their size.

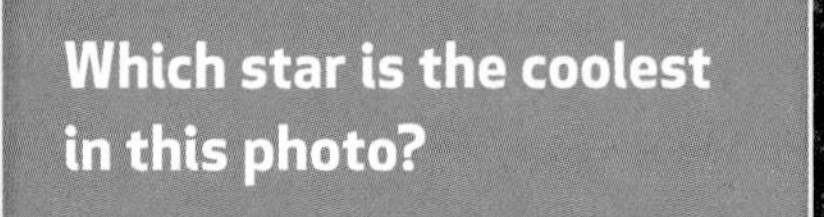

Which star is the coolest in this photo?

Betelgeuse is one of the biggest stars in the sky. But another large star, Rigel, is brighter because its temperature is hotter. Look at the stars shown in the chart. Compare the color and temperature of the sun with the other stars in the chart.

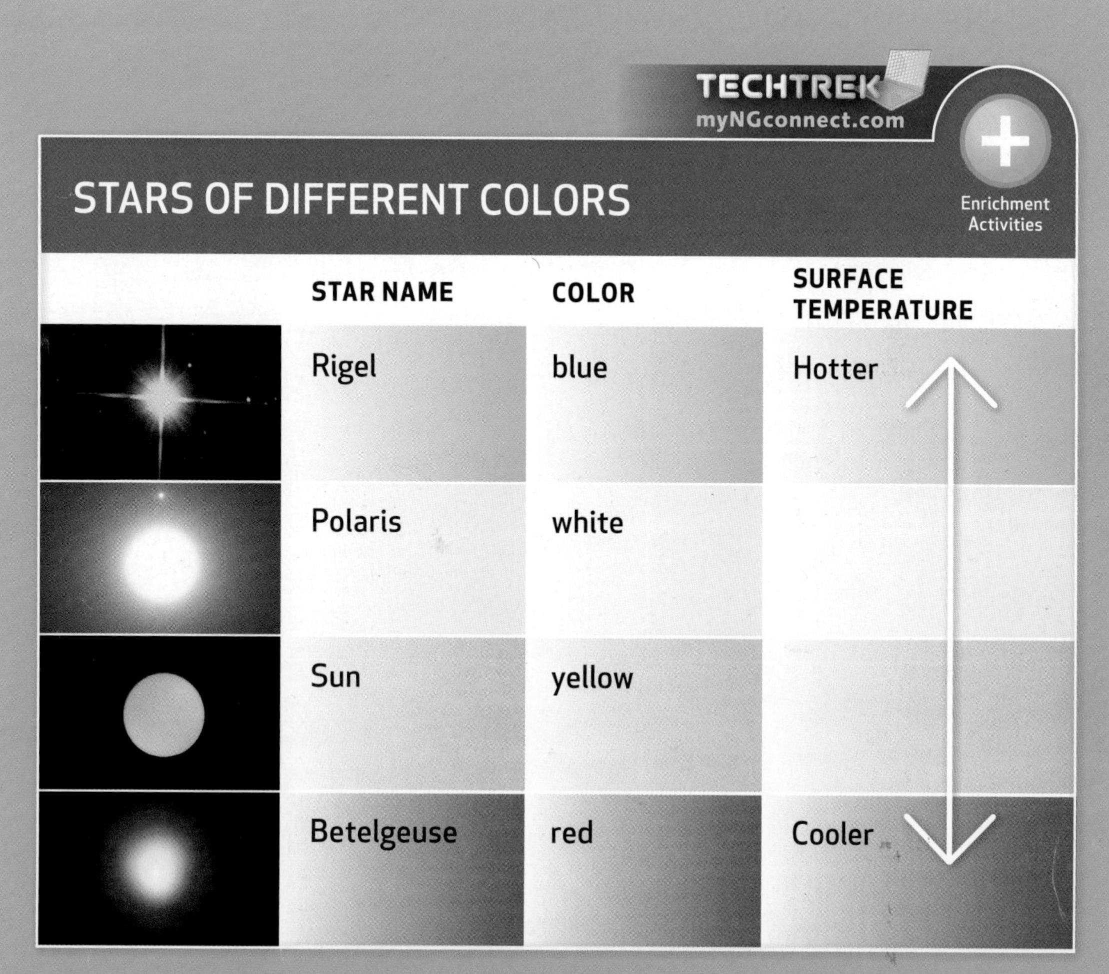

STARS OF DIFFERENT COLORS

	STAR NAME	COLOR	SURFACE TEMPERATURE
	Rigel	blue	Hotter
	Polaris	white	
	Sun	yellow	
	Betelgeuse	red	Cooler

Before You Move On

1. What two properties play a role in how bright a star is?
2. What is the difference between red stars and blue stars?
3. **Evaluate** If a star looks brighter than another star, does that always mean it is bigger than the other star? Why or why not?

Observing Stars

Scientists have used **telescopes** for hundreds of years. A telescope is a tool that magnifies objects. It makes faraway objects look closer and bigger. You can see more detail and learn more about the objects. A telescope collects more light than your eyes can. So with a telescope, you can see more stars than with your eyes alone.

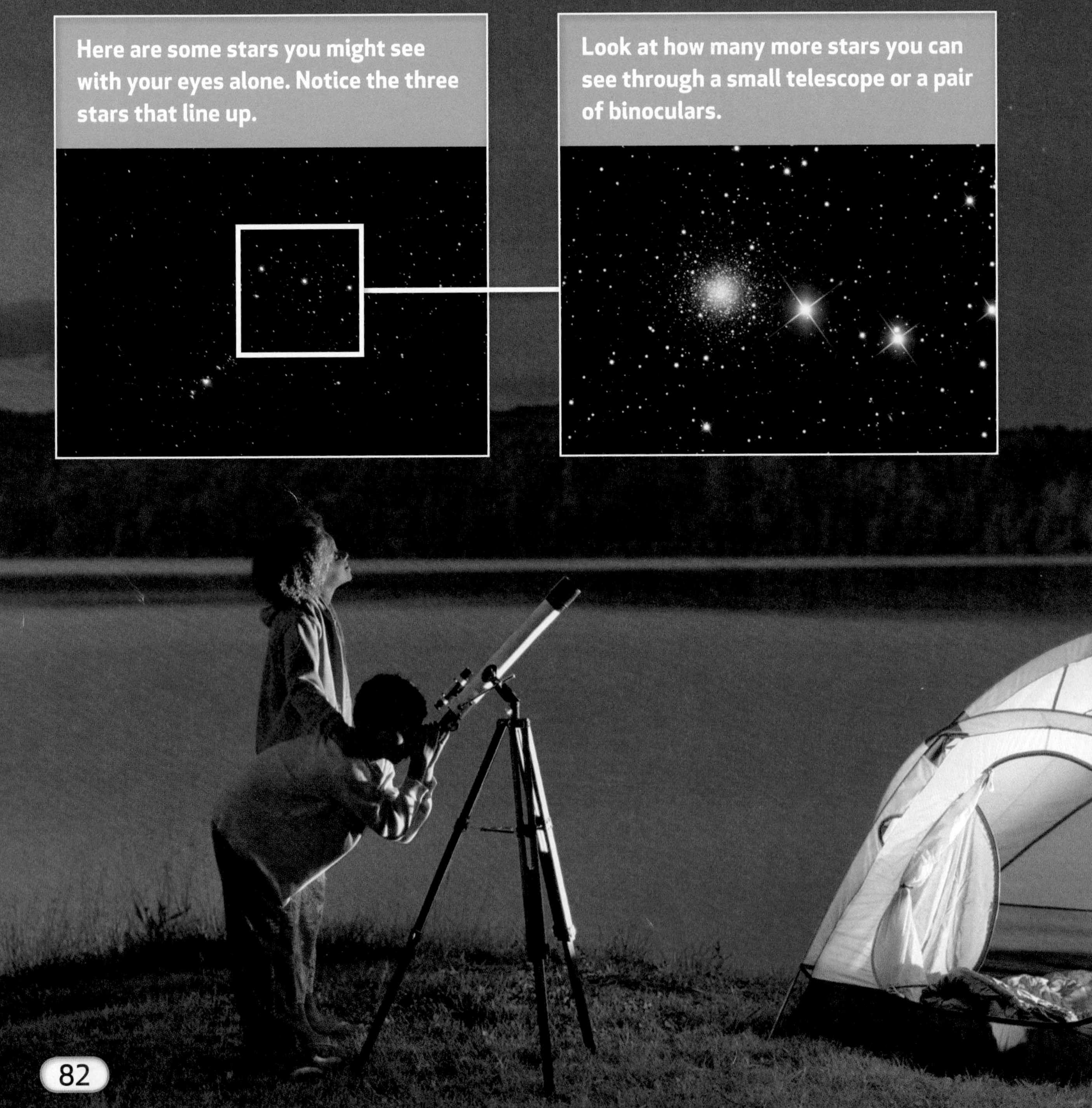

Here are some stars you might see with your eyes alone. Notice the three stars that line up.

Look at how many more stars you can see through a small telescope or a pair of binoculars.

Early telescopes were not very powerful. But telescopes today are bigger and much more powerful. Scientists use them to see deeper and deeper into space. They have discovered more stars than ever before. In the future, new powerful telescopes will help scientists learn even more about stars.

Science in a Snap! Seeing Better

Look at a coin. What can you observe about it?

Look at the coin through a hand lens. Do you notice anything you didn't see before?

How does the hand lens help you observe the coin? How is this like looking at the night sky through a telescope?

The stars can be seen as night falls over Caspian Lake, Vermont.

At the Gemini North Observatory in Hawaii, scientists use a large telescope to observe the night sky. They use the telescope, computers, and other tools to watch and take pictures of stars. Gemini North was built on top of a mountain so scientists can see the whole sky.

With a powerful telescope, scientists can see the colorful light from gas and dust in space.

The Gemini North Observatory sits on top of a mountain. The observatory is 4200 meters (14,000 feet) above sea level. That's higher than most clouds.

Using Gemini North's telescope, scientists watch many different stars. They observe stars that have just formed. They observe stars that have faded away or exploded.

TECHTREK
myNGconnect.com

Digital Library

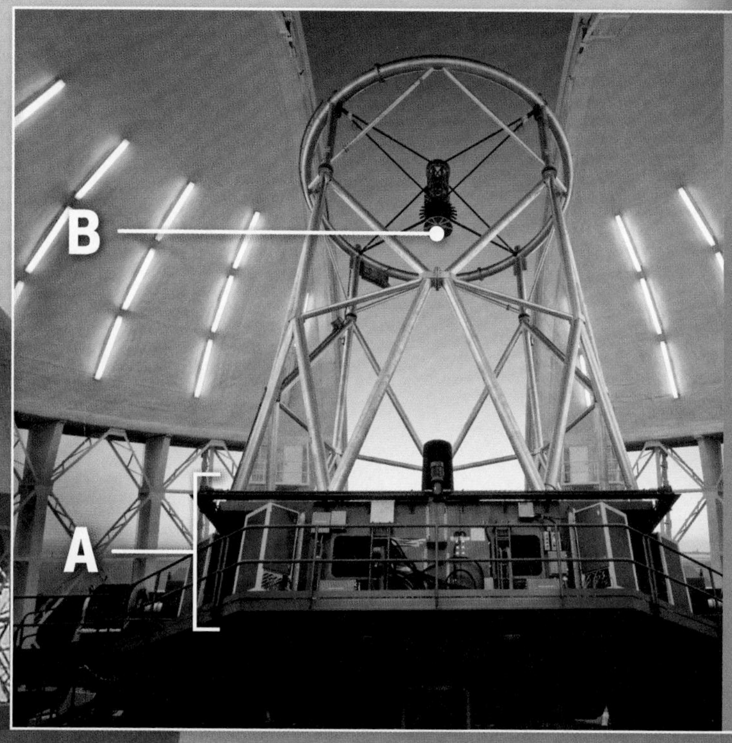

TELESCOPE

This telescope is inside the Gemini North Observatory. Special mirrors focus and reflect light.

A. The first mirror is at the base of the telescope. It is over 8 meters (27 feet) wide.

B. Light from the first mirror is reflected up to the second mirror.

A camera inside the telescope records images.

Before You Move On

1. What does a telescope do?
2. What might you see through a telescope that you cannot see with just your eyes?
3. **Infer** Why would the Gemini North Observatory not be as useful if it were built in a city?

ANCIENT LIGHT

Proxima Centauri is Earth's second-closest star. It is four light-years away. So if this star exploded today, we wouldn't see the explosion for another four years! How can that be? You see an object only because light travels from the object to your eyes. Stars are so far away that their light takes years to reach us. Scientists use a unit called a light-year to measure how far stars are from Earth. A light-year is the distance light moves in a year. A light-year is about 9.5 trillion kilometers (about 6 trillion miles).

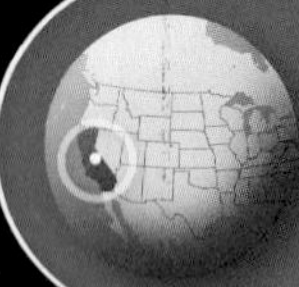

People stargaze in Yosemite National Park in California.

Sunlight takes only eight minutes to reach Earth. When explosions happen on the sun, we see them eight minutes later. Compare eight minutes to 800 years—that's how long it takes light from a star like Rigel to reach Earth. Rigel is 800 light-years away. Most stars are thousands or millions of light-years away.

This disc-shaped group of stars is almost 3 million light-years away. If scientists would observe something happening there, the event would have happened 3 million years ago.

This picture was taken over the rocks of Joshua Tree National Park in California. It shows the blue-white star Sirius. It is the brightest star seen in the night sky. It is nearly 9 light years away.

Conclusion

Stars are huge balls of hot gases. They are so far away that they look like points of light in the night sky. Stars are not all the same. Some are closer to Earth than others. Some are bigger and brighter than others. Stars also have different surface temperatures, which make them different colors. A star's brightness depends on its size and surface temperature. You can see many more stars with a telescope than you can with your eyes alone.

Big Idea You can observe several properties of stars including brightness, size, and color.

Brightness

Size

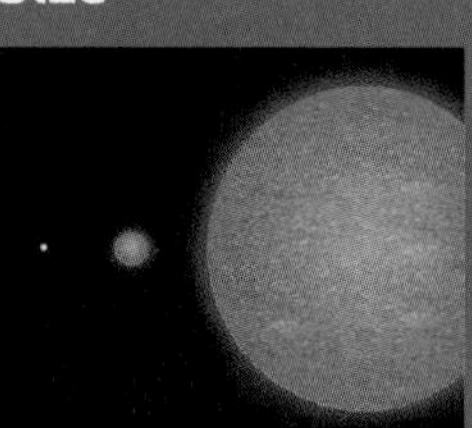

Temperature

Color

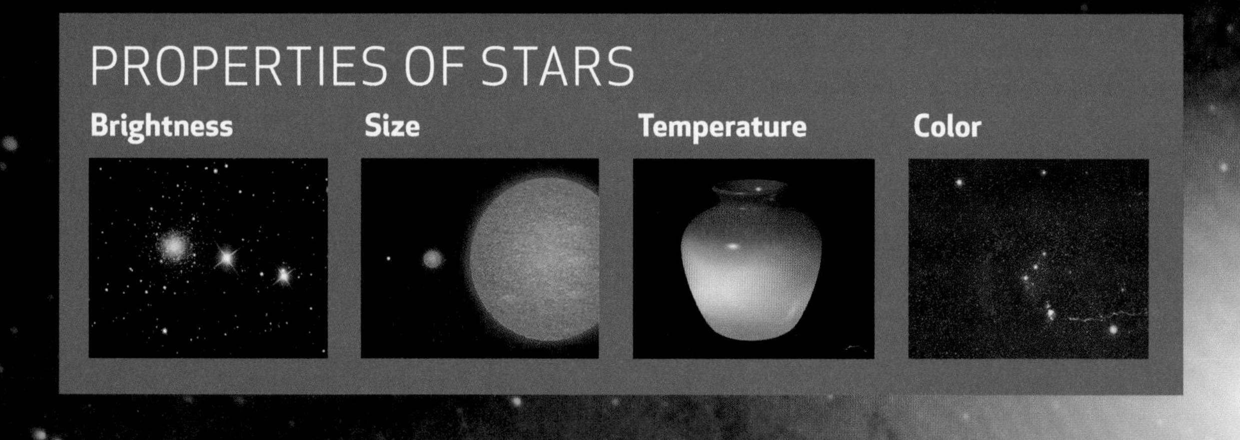

Vocabulary Review

Match the following terms with the correct definition.

A. star
B. brightness
C. telescope
D. property

1. A glowing ball of hot gases
2. A tool that magnifies objects and makes them look closer and bigger
3. The amount of light that reaches your eye from an object such as a star
4. Something about an object that you can observe

Big Idea Review

1. **Organize** Order the star colors from coolest to hottest: blue, red, white, yellow.
2. **Describe** How does a telescope work?
3. **Explain** Why do stars seem so small when you look at them with your eyes alone?
4. **Compare and Contrast** How are hot stars and cool stars alike and different?
5. **Make Judgements** Why do you think it is better to build an observatory high on a mountain rather than lower down?
6. **Generalize** You have learned about the properties of stars. Use these properties to describe the sun.

Write About the Night Sky

Cause and Effect The sun is a star you can see during the day. Write a few sentences to explain why you can't see other stars during the day.

CHAPTER 3

EARTH SCIENCE EXPERT: ASTRONOMER

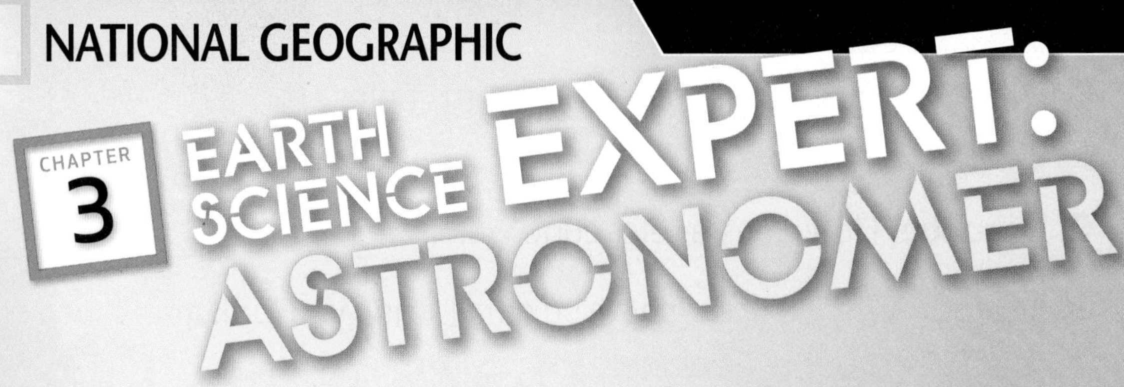

Want to study the stars and beyond? Astronomer Jason Kalirai talks about outer space.

What does an astronomer do?

I use large telescopes to take pictures of planets, stars, and galaxies in the universe. I use telescopes on Earth and in space. I then study the pictures to learn more about these objects.

What's the favorite part of your job?

My favorite part is when I first look at a new picture of a star or galaxy. It's exciting to see something for the first time, especially if something unexpected is discovered.

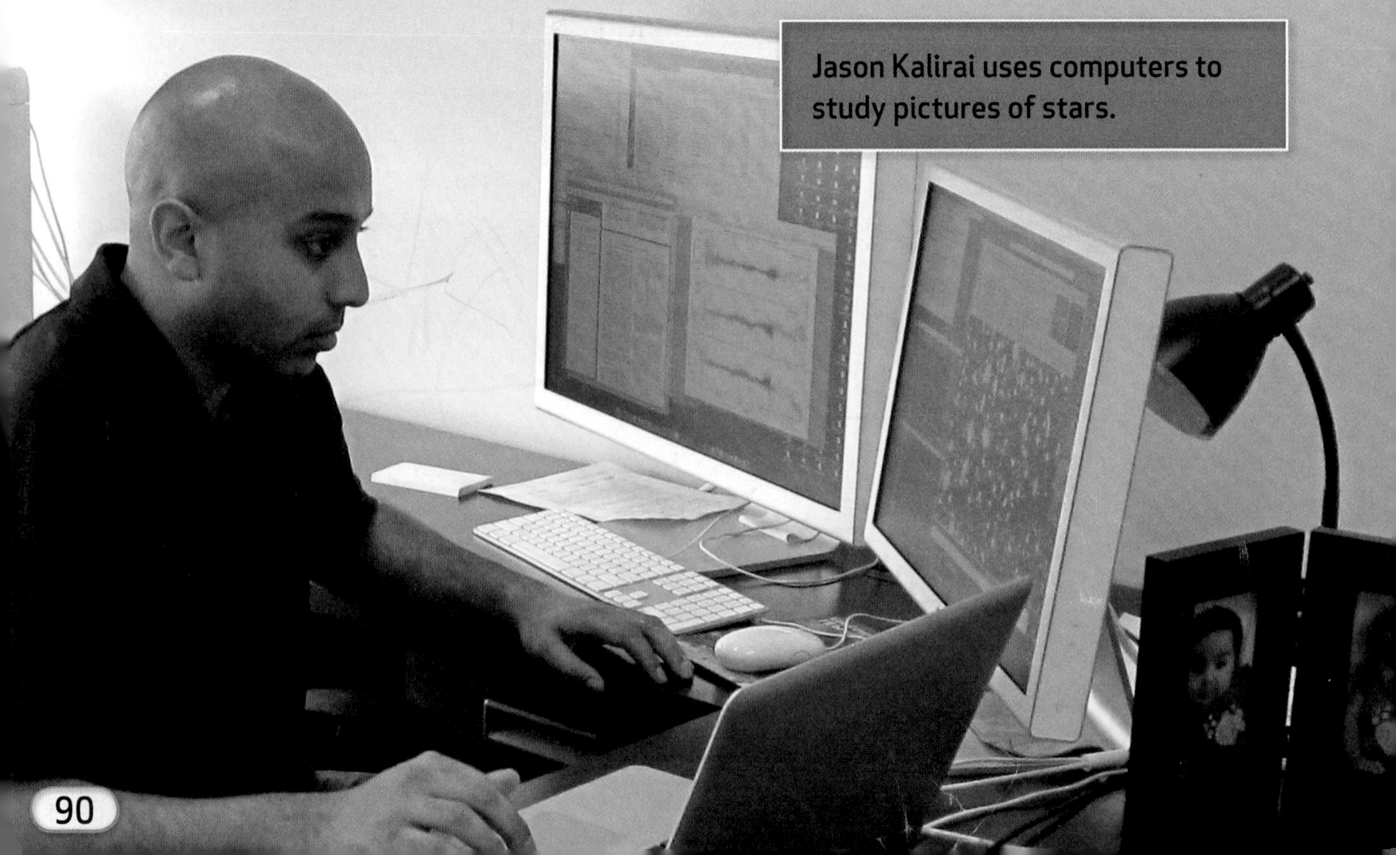

Jason Kalirai uses computers to study pictures of stars.

How did you become interested in astronomy?

As a kid, I loved looking at the night sky. I learned all I could about the stars and planets.

What's a typical day like for you?

I usually go to work from 9:00 a.m. to 5:00 p.m., like most people. At work, I spend time discussing research with fellow astronomers and sharing observations. I also spend a lot of time traveling. I've been all over the U.S., including Hawaii. I've been to other countries too, including Australia, New Zealand, India, Japan, Egypt, and many places in Europe. While visiting these places, I obtain new observations and meet new astronomers.

Why should students become astronomers?

I find my job to be the coolest one in the world, and I want others to see why. We need new, young scientists who will think creatively. That means asking questions no one has asked before. It means observing something and explaining it in new ways.

Jason zeroes in on clusters of stars to learn more about them.

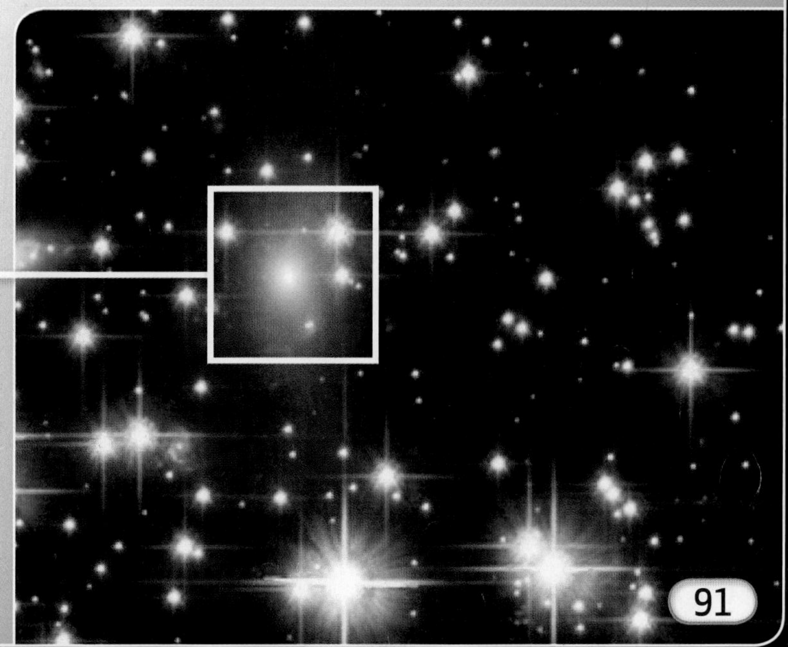

NATIONAL GEOGRAPHIC

The Hubble Space Telescope: Super Stargazer

From Earth, it is hard to see many of the glowing balls of hot gases called **stars**. This is because the moving gases in the air around Earth can block starlight. To get a better view of stars and space, scientists launched the Hubble Space **Telescope** in 1990.

The Hubble Space Telescope flies 569 kilometers (353 miles) above Earth.

stars

A **star** is a glowing ball of hot gases.

telescope

A **telescope** is a tool that collects light from objects and magnifies them.

The space telescope flies high above Earth. It circles our planet once every 97 minutes. All the while, it points into space and takes pictures. The telescope sends these pictures back to Earth. Studying the photos has taught scientists a lot about the universe. Some of Hubble's photos show colorful clouds of gas and dust in space. Such a cloud is called a nebula. Stars are born in a nebula.

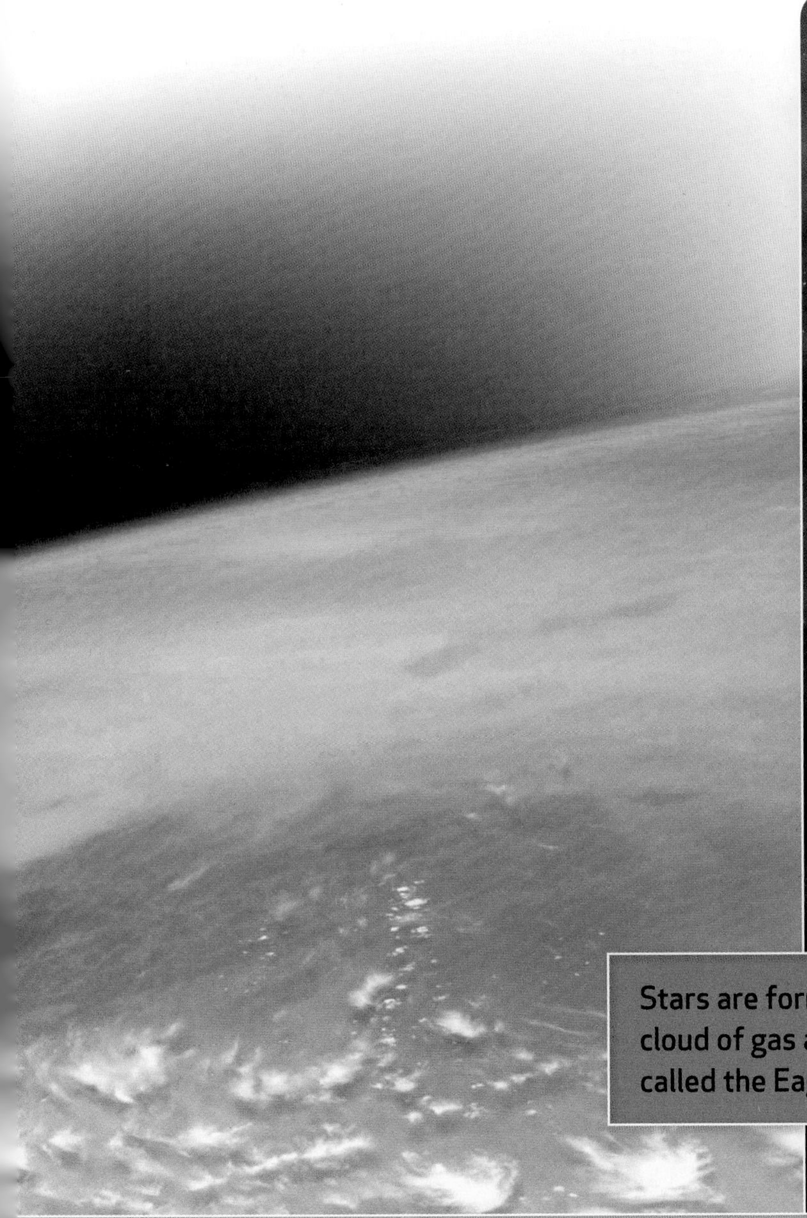

Stars are forming in this cloud of gas and dust called the Eagle Nebula.

A Star Is Born

Stars form when bits of gas and dust come together inside a nebula. A force called gravity pulls together gas and dust. This material forms a clump, which grows larger and larger. As the clump grows, so does its gravity.

TECHTREK
myNGconnect.com
Digital Library

Many stars form in the Swan Nebula. This nebula is 5,000 light-years from Earth.

Gravity pulls more and more gas and dust into the clump. It also pulls the clump into a tight ball. As the ball gets tighter and tighter, its **properties** change. The temperature rises. If the ball becomes hot enough, it starts shining. At that point, a star is born. Its **brightness** might be seen from Earth. All of this takes millions of years.

HOW DO STARS FORM?

Gravity begins to pull gas and dust together into a clump.

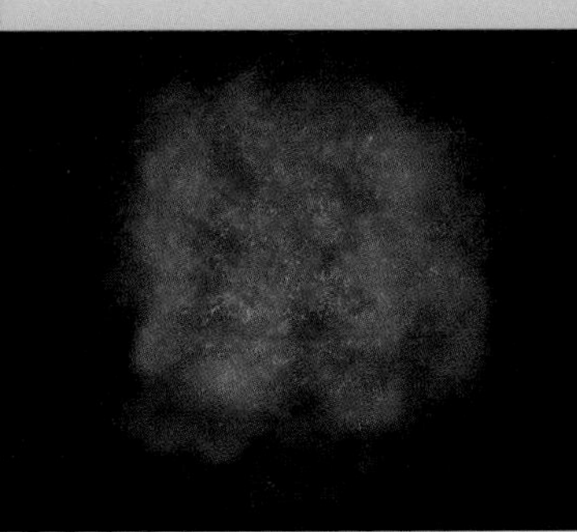

The clump forms a ball called a protostar.

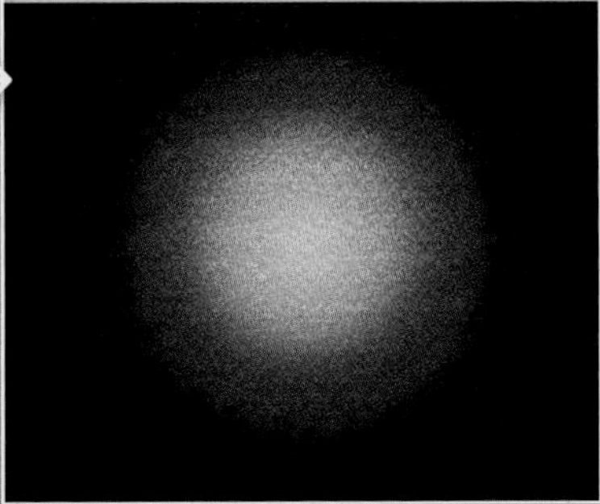

Gas and dust in the protostar heat up enough to give off a lot of energy. A star is born.

property

A **property** is something about an object that you can observe.

brightness

Brightness is the amount of light that reaches your eye from an object such as a star.

Dying Stars

Photographs from the Hubble Space Telescope are also helping scientists learn about the end of a star's life. No star lasts forever, and that includes our sun. It is five billion years old. But don't worry. It should stay as it is for another five billion years. Then it will start to cool down, turn red, and grow larger.

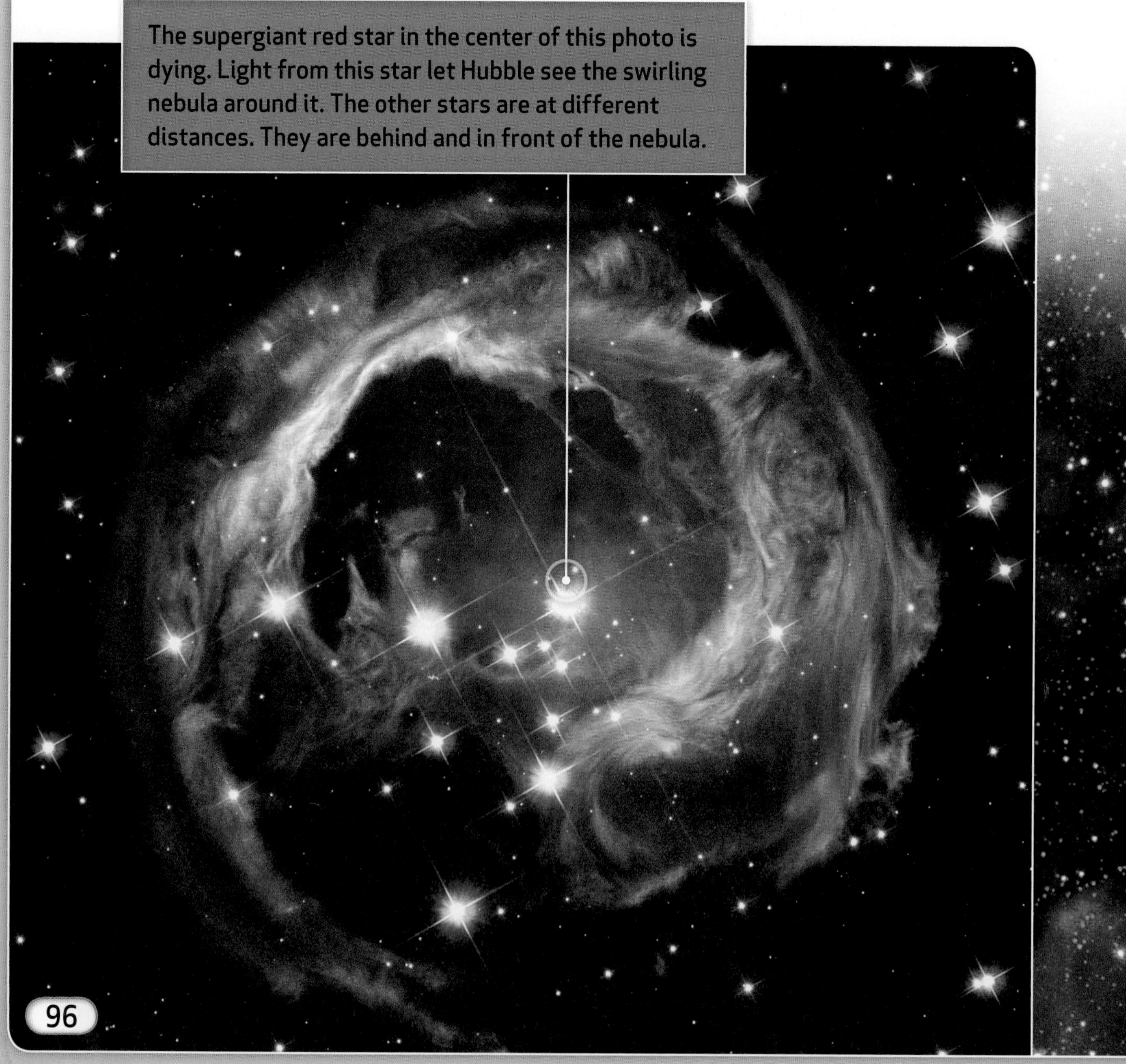

The supergiant red star in the center of this photo is dying. Light from this star let Hubble see the swirling nebula around it. The other stars are at different distances. They are behind and in front of the nebula.

Near the end of its life, the sun might even grow large enough to touch Earth! A star that size is called a red giant. After a while, the sun will stop growing and shed its outer layers. Photos from Hubble have shown that these layers float away gently, like a puff of smoke. They form a nebula around the dying star. Later, it will fade to black and die. But new stars will form in the nebula left behind.

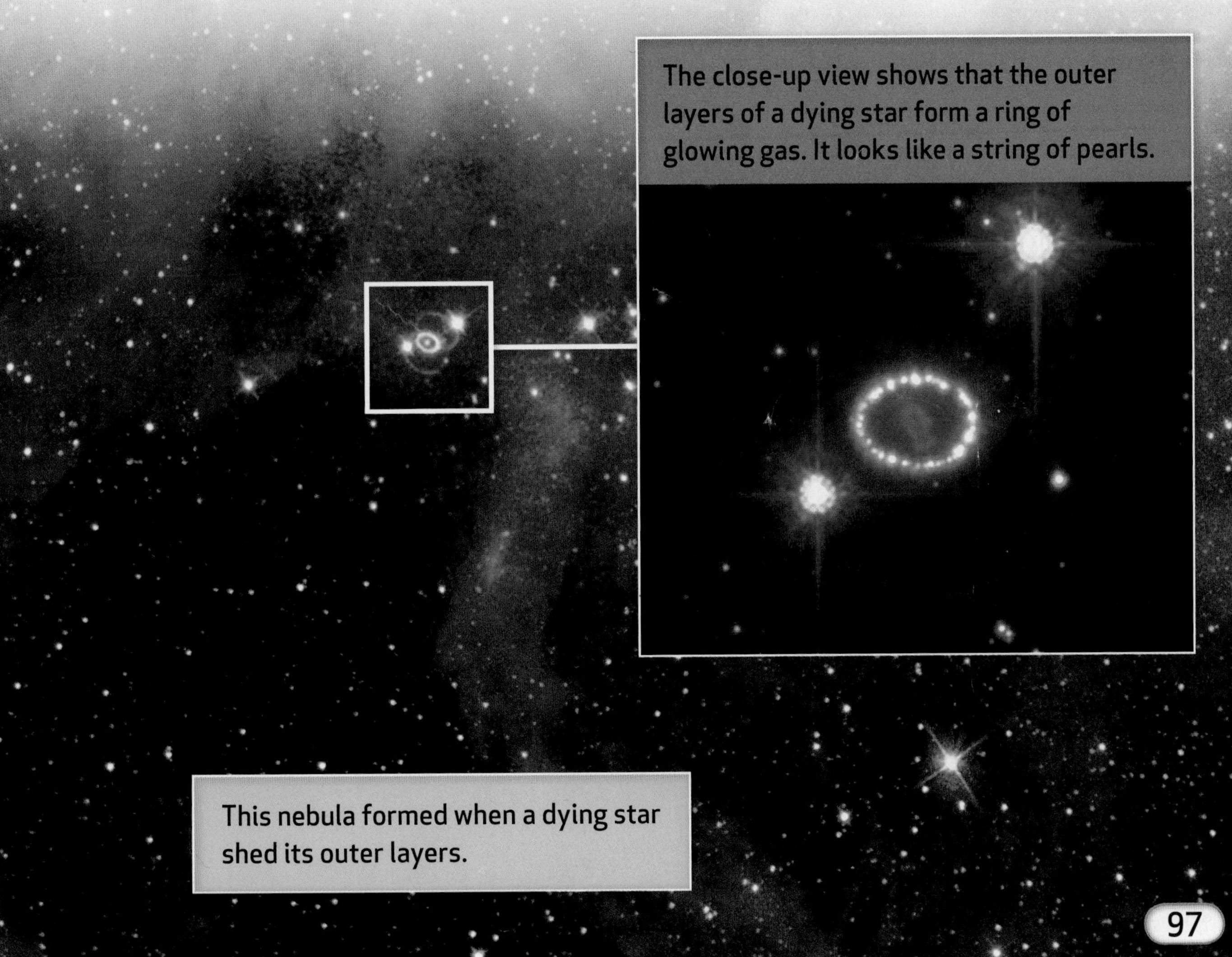

The close-up view shows that the outer layers of a dying star form a ring of glowing gas. It looks like a string of pearls.

This nebula formed when a dying star shed its outer layers.

Looking Ahead

The Hubble Space Telescope has taught us about the lives of stars and much more. But the Hubble can only do so much. Soon a new telescope will join the Hubble high above Earth. In 2014, scientists plan to launch the James Webb Space Telescope. The Webb telescope will let scientists look even deeper into space.

This picture shows what the James Webb Space Telescope will look like in space.

For 20 years, the Hubble Space Telescope has taken pictures of faraway stars and galaxies. Scientists have studied photos taken by Hubble to learn about the life cycle of stars. In the future, Hubble and the James Webb Space Telescope will tell us even more about our amazing universe.

In the Orion Nebula, several white-hot stars have already formed. The swirling clouds of gas and dust will form even more stars.

NATIONAL GEOGRAPHIC

CHAPTER 3 SHARE AND COMPARE

Turn and Talk How are stars born and how do they die? Form a complete answer to this question together with a partner.

Read Select two pages in this section. Practice reading the pages. Then read them aloud to a partner. Talk about why the pages are interesting.

my SCIENCE notebook **Write** Write a conclusion that tells the important ideas you have learned about the Hubble Space Telescope. State what you think is the Big Idea of this section. Share what you wrote with a classmate. Compare your conclusions.

my SCIENCE notebook **Draw** Form groups of three. Have each person draw a star at different points in its life. Each person should label their drawings and write captions to explain what is happening in the drawings. Present your drawings and explanations to the class.

After reading Chapter 4, you will be able to:

- Recognize that rocks and soil are Earth materials. **ROCKS AND SOIL**
- Recognize that rocks and soil are made of minerals. **MINERALS**
- Recognize that rocks can have different properties. **COMPARE ROCKS**
- Identify and describe the properties of soil. **SOIL**
- Science in a Snap! Recognize that rocks can have different properties. **COMPARE ROCKS**

CHAPTER

4

WHAT CAN YOU OBSERVE ABOUT

Have you ever dug a hole in the ground?

If you have, you probably found soil and rocks. Soil and rocks are two of the materials found on Earth. If you look closely, you will find that not all rocks and soil are the same. How can you describe the rocks and soil where you live?

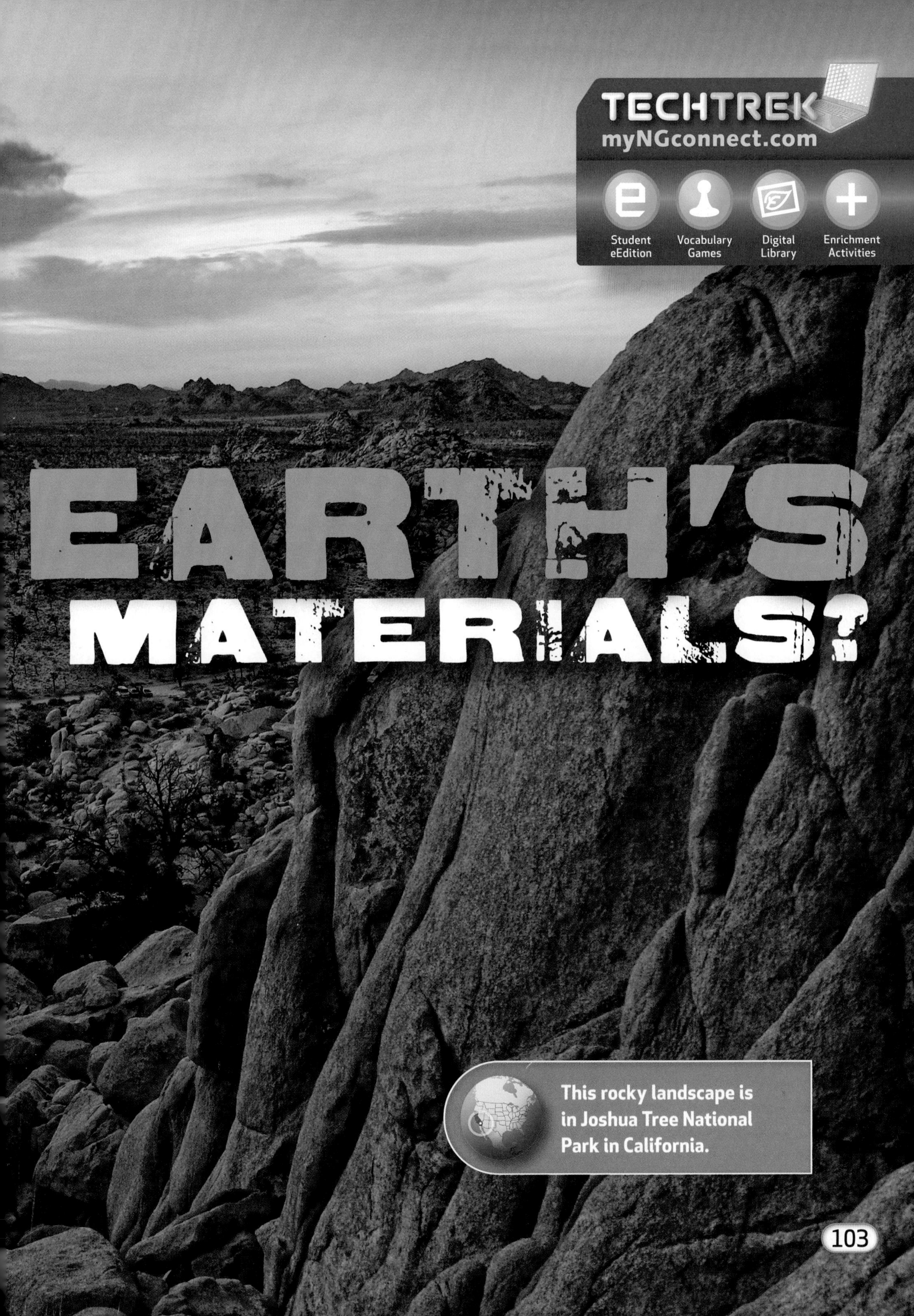

This rocky landscape is in Joshua Tree National Park in California.

SCIENCE VOCABULARY

soil (SOIL)

Soil is a layer of loose materials on Earth's surface that is made up of bits of rocks, humus, air, and water. (p. 107)

Some soil is sandy, and some is sticky like clay.

mineral (MIN-ur-ul)

Minerals are solid nonliving materials found in nature. (p. 108)

These minerals are different from each other in many ways.

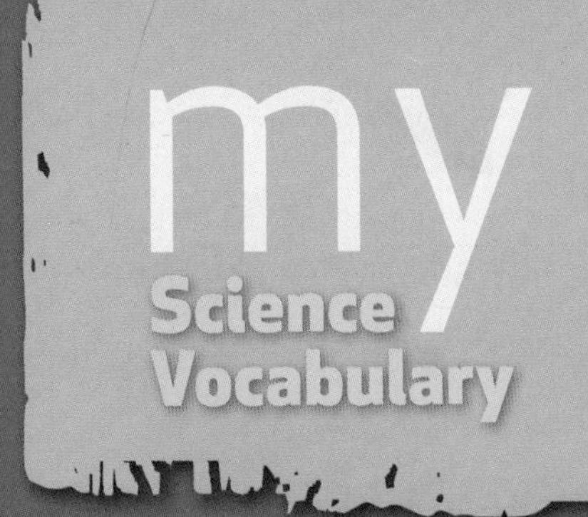

grain (GRĀN)	**mineral** (MIN-ur-ul)
humus (HYŪ-mus)	**soil** (SOY-il)

Vocabulary Games

grain (GRĀN)

A **grain** is a piece that makes up rock or soil. (p. 109)

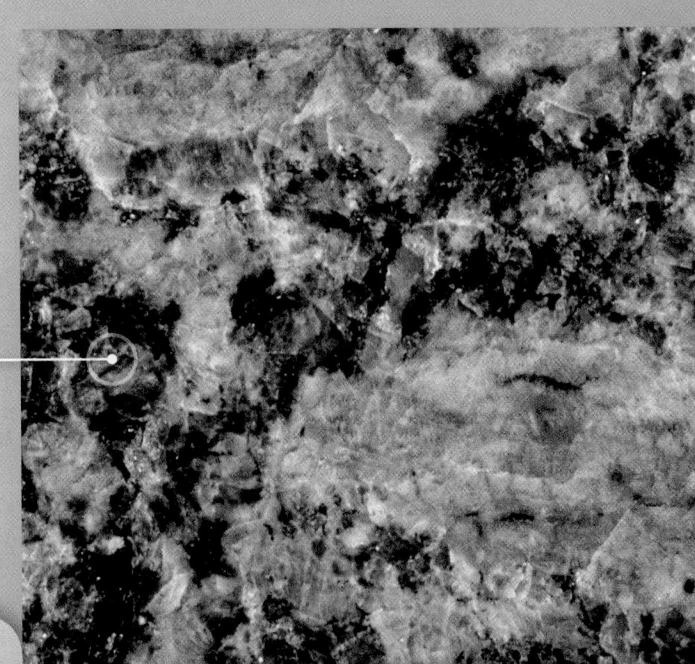

You can see the grains in this rock without a hand lens.

humus (HYŪ-mus)

Humus is a part of soil made of bits of decayed plants and animals. (p. 114)

This earth material has a lot of humus in it.

Rocks and Soil

Earth's land is made mostly of rocks. These materials are not the same everywhere on Earth. You can see huge rocks called boulders that are big enough to sit on. At the beach you run through tiny pieces of rocks on your way to the water. Sand is also an Earth material.

Earth's land is also made of **soil**. This is a layer of loose materials on Earth's surface that is made up of bits of rocks, decayed matter, air, and water. Some soil contains clay. Clay is made of the tiniest rock pieces. It feels sticky when wet.

TECHTREK myNGconnect.com

Digital Library

SIZES OF ROCKS

BOULDERS
Boulders are large rocks.

GRAVEL
Gravel is made of smaller rocks you can hold in your hand.

SAND
Sand is tiny pieces of rock.

Before You Move On

1. What is soil?
2. Name four sizes of rocks. Put them in order.
3. **Compare and Contrast** Describe the rocks on the beach in the big photo.

Minerals

Most rocks and soil are made of **minerals**. These are solid nonliving materials found in nature. You use minerals every day. Salt is a mineral called halite. The tip of your pencil is a mineral called graphite. The coins in your pocket may contain the mineral copper. The jewelry you wear may be made of minerals. Gold, silver, and platinum are all minerals.

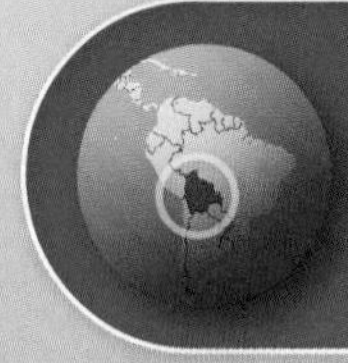

These men are loading salt onto a truck from a dried-up lake in Bolivia.

Soil and rocks may contain minerals. In fact, most rocks are a mixture of different minerals. A **grain** is a small piece that makes up rock or soil. Some grains are too small to see without a hand lens. Other grains are bigger. Look at the granite. This rock has grains you can see. It contains the minerals quartz, feldspar, and mica.

Use the labels to see what the different minerals in granite look like. Point to other parts of the photo that show each mineral.

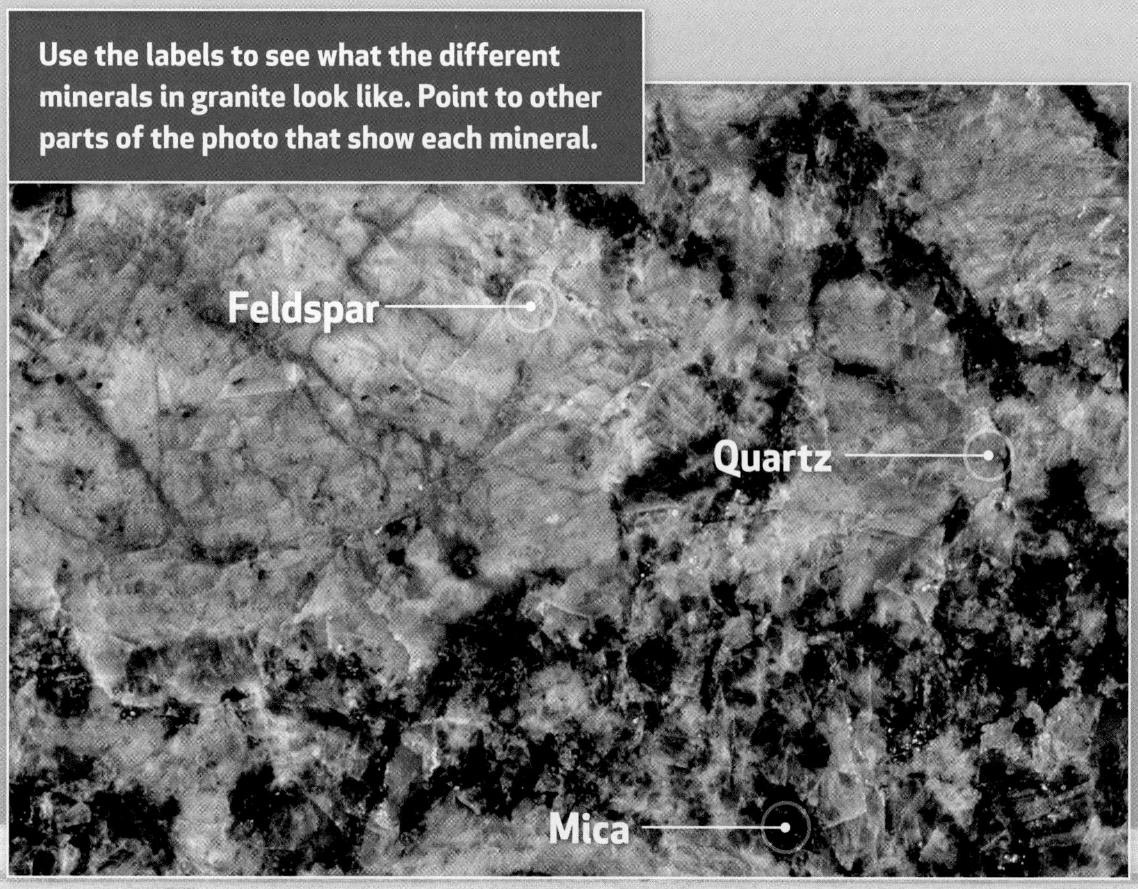

How can you tell minerals apart? They have different properties. A property is what you can observe about something with your senses.

Color is a property you observe with sight. But you cannot use color alone to name a mineral. Some minerals come in different colors. Also, many different minerals have the same color.

The mineral azurite is always blue.
The mineral fluorite can be purple, yellow, green, or blue.
The mineral malachite is always green.

You can tell more about a mineral by its streak. Rub a mineral on a tile. The powder left on the tile is called a streak. A mineral's streak is always the same color.

Hardness is another property of minerals. The hardest mineral on Earth is diamond. Nothing scratches diamond. Talc is the softest mineral. Talc is so soft you can scratch it with your finger nail.

Hematite can be different colors. But its streak is always red.

This photo was taken through a microscope. The blue pieces are tiny diamonds on a drilling tool. The diamonds on this drill can make a hole in almost anything.

Before You Move On

1. Define minerals.
2. Compare color and streak. Which is better for telling minerals apart?
3. **Apply** The mineral copper will not scratch glass. Glass will scratch copper. Which is harder, copper or glass?

Compare Rocks

Most rocks are made of minerals. Rocks have different properties because of the minerals that are in them. The way rocks form also affects their properties.

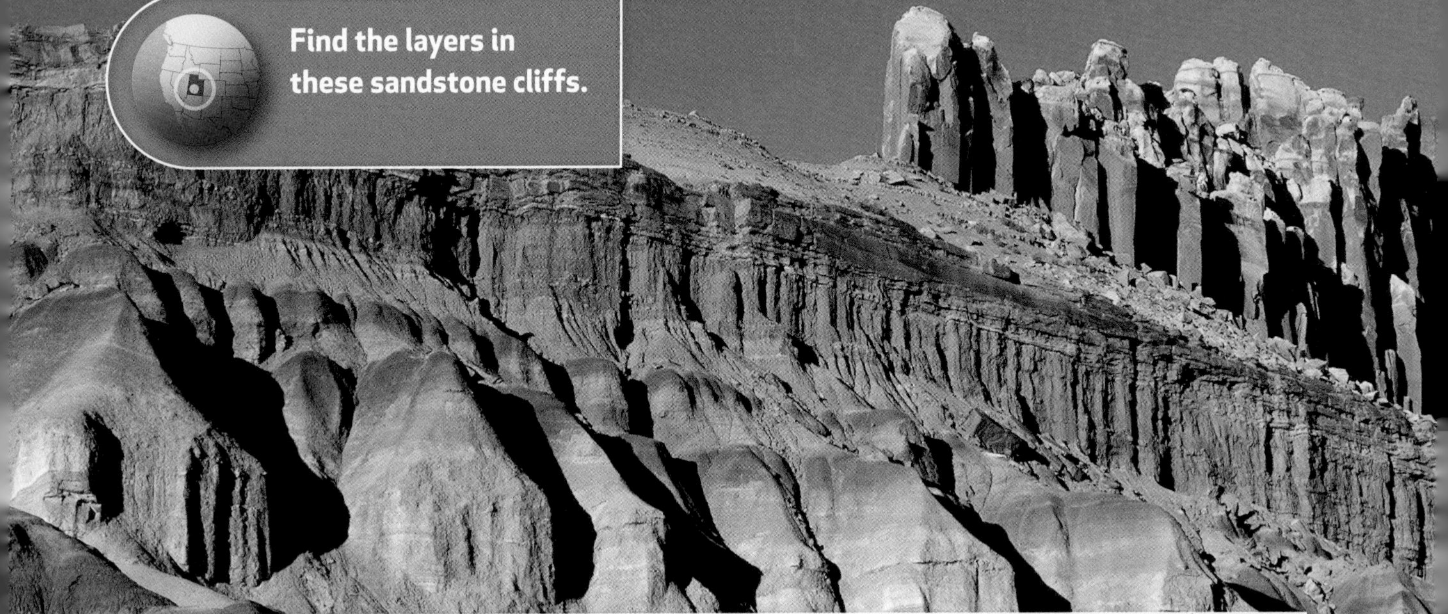

Find the layers in these sandstone cliffs.

Science in a Snap! Rock Properties

Use pieces of sandstone and pumice.

Put the rocks in the water. Observe them.

What properties did you observe?

Some rocks form in layers. Sandstone forms from buried layers of sand mixed with minerals. Pressure squeezes the layers together. Over time, the layers harden into rock. This can take thousands or millions of years.

Texture is another property. A rock's texture comes from the size of its grains. Rocks with larger grains are more rough or coarse. Rocks with small grains are smooth or glassy.

Enrichment Activities

COMPARING SOME PROPERTIES OF ROCKS

COLOR gray, red	 **gray**	 **red**
TEXTURE coarse, glassy	 **coarse**	 **glassy**
LAYERS layered, not layered	 **layered**	 **not layered**

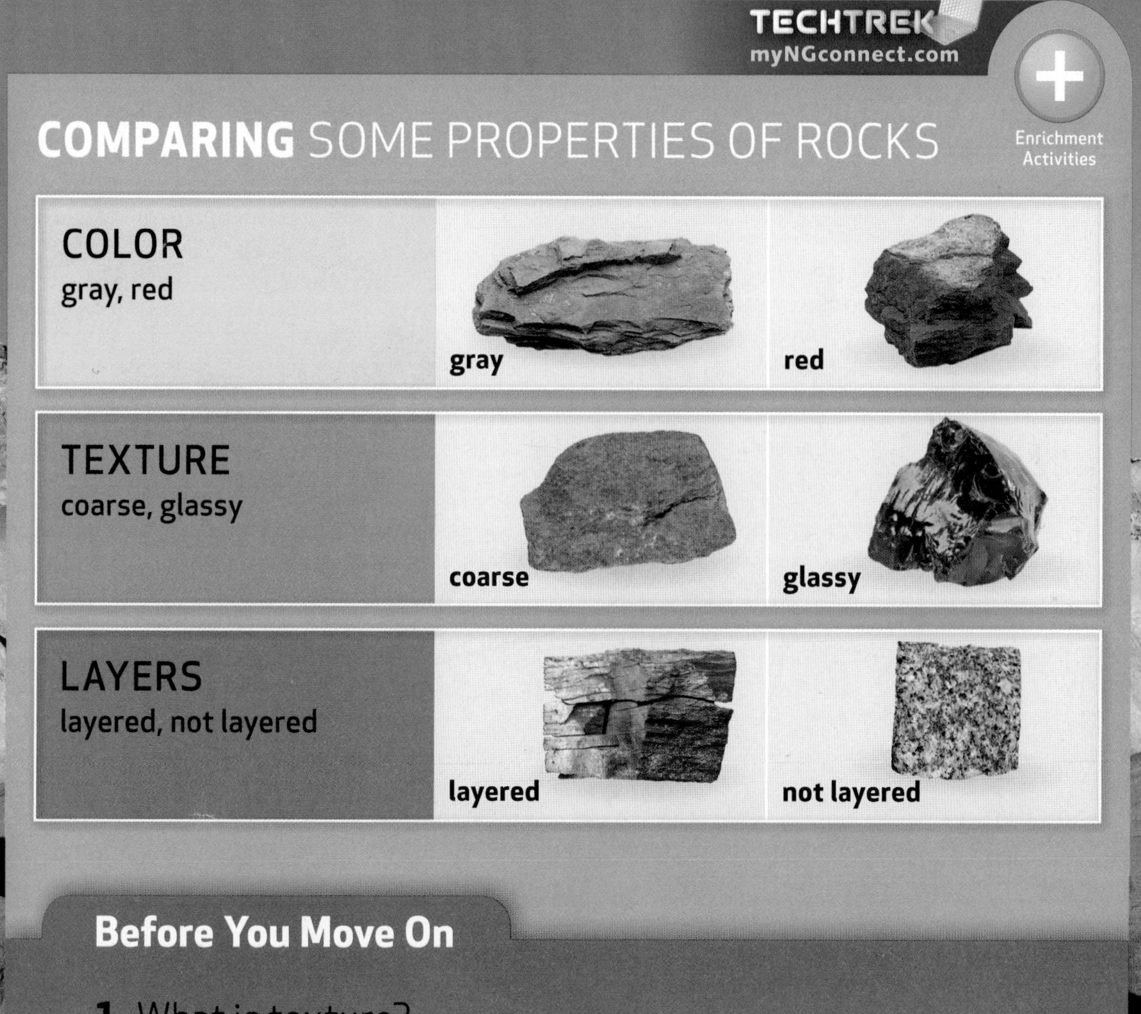

Before You Move On

1. What is texture?
2. List two properties of rocks.
3. **Apply** Pick a rock from the chart. Name one other property you can observe about it.

Soil

Have you ever planted a garden? Then you were digging in soil. Soil forms as rocks break down into smaller and smaller pieces. The pieces may be the size of sand or clay. These pieces of rock then mix with air and water. To be soil, the mixture of rock pieces, air, and water also needs **humus**. Humus is bits of decayed plants and animals.

Soil is important to people and most living things. Plants grow in soil. You eat the plants, and so do animals. Soil is also used to make pottery and bricks.

Soil is an important resource, but it takes a long time to form. That is why people try to conserve soil. They try to keep it from blowing away in the wind or washing away in the rain.

Wind blows soil away. So some farmers plant a line of trees to block the wind.

Types of Soil Soil comes in three main types. Look at the types of soil pictured here. What can you observe about sandy soil, clay soil, and loam?

One of the differences between soils is their texture. Just like with rocks, texture comes from the size of the grains. Sandy soil has the largest grains. Clay soil has the smallest.

SANDY

Sandy soil contains more sand. It is usually brown.

- **It has larger grains.**
- **It drains quickly and does not hold water well.**
- **It contains a lot of air.**
- **It has very little humus.**

A soil's texture determines how easily water drains, or trickles through the soil. If water drains quickly, the soil will be very dry. If water drains slowly, the soil will be wetter.

People use soil for many things. But one thing soil is used for all over the world is growing plants. No one kind of soil is best for all plants. But almost all plants depend on soil.

CLAY

Clay soil contains more clay. It is often red.

- **It has very small grains.**
- **It drains slowly and holds water well.**
- **It contains very little air.**
- **It has some humus.**

LOAM

Loam contains some sand and clay, but mostly humus. It is often dark brown.

- **It has different grain sizes.**
- **It holds some water but drains well.**
- **It contains the right amount of air.**
- **It has a lot of humus.**

Before You Move On

1. List the four parts of soil.
2. Explain how soil is important to people.
3. **Apply** Suppose you have two piles of soil—sandy and clay. You also have a bag of humus. Explain how you can use these to make loam.

NATIONAL GEOGRAPHIC

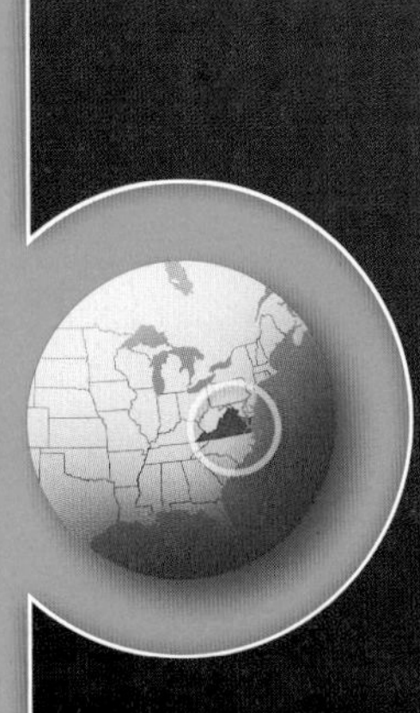

STONE TOOLS FROM THE PAST

Think about the tools people use today. Most tools are probably made from metal parts. Over 12,000 years ago, people made tools from rocks such as flint. They made flint into arrow heads and spear points for hunting. They even used flint to make sparks to start fires. With fire, people could keep warm, cook food, and provide light.

Scientists found these stone tools at a dig in Virginia. How do you think people used these tools?

A scientist is making an arrowhead out of flint. He does this by chipping away at the flint with another rock.

Some scientists who study ancient humans have relearned how to make stone tools. These scientists needed to know how people in ancient times made them. Knowing how to make their tools is one of the best ways to understand how the people lived.

Stone tools have lasted for thousands of years. Scientists dig them up and carefully observe them. You can also learn how people in the ancient past lived by studying their tools.

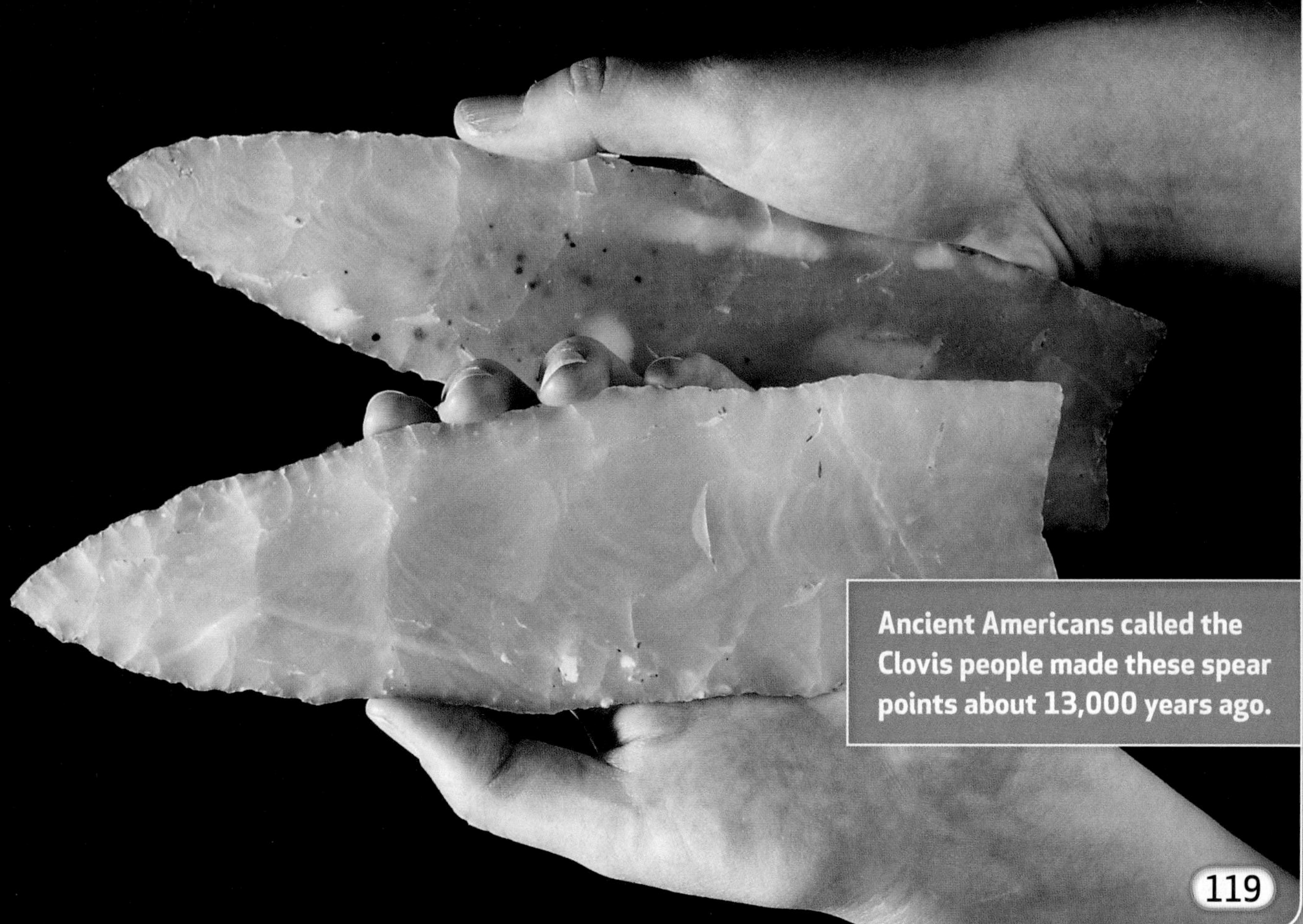

Ancient Americans called the Clovis people made these spear points about 13,000 years ago.

Conclusion

Rocks and soil are Earth materials. Rocks can be different sizes. Rocks and soil are made mostly of minerals. You can observe minerals' properties, such as color, streak, and hardness. Color, texture, and layers are some properties of rocks and soil. Soil forms as rocks break down into tiny pieces that mix with water, air, and humus.

Big Idea You can observe the properties of Earth materials.

TYPES OF EARTH MATERIALS

Vocabulary Review

Match the following terms with the correct definition.

A. humus
B. soil
C. minerals
D. grain

1. A small piece that makes up rock or soil
2. A part of soil made of bits of decayed plants and animals
3. Solid nonliving materials found in nature
4. A layer of loose materials on Earth's surface that is made up of bits of rocks, humus, air, and water

Big Idea Review

1. **Recall** What is one way to identify a mineral?

2. **Identify** What Earth material are most rocks made of?

3. **Classify** Classify the following as a mineral, rock, or soil: granite, iron, copper, loam, sandstone.

4. **Sequence** How do the layers in sandstone form?

5. **Draw Conclusions** One pot of soil holds too much water for its plant. Another pot holds just the right amount of water. The two soils are different colors. What conclusions can you draw about them?

6. **Generalize** What do all types of soil have in common? Explain.

Write About Rock Properties

Analyze What are the properties of this rock? Analyze the rock using the properties you learned about in the chapter.

NATIONAL GEOGRAPHIC

CHAPTER 4

EARTH SCIENCE EXPERT: ECOLOGIST

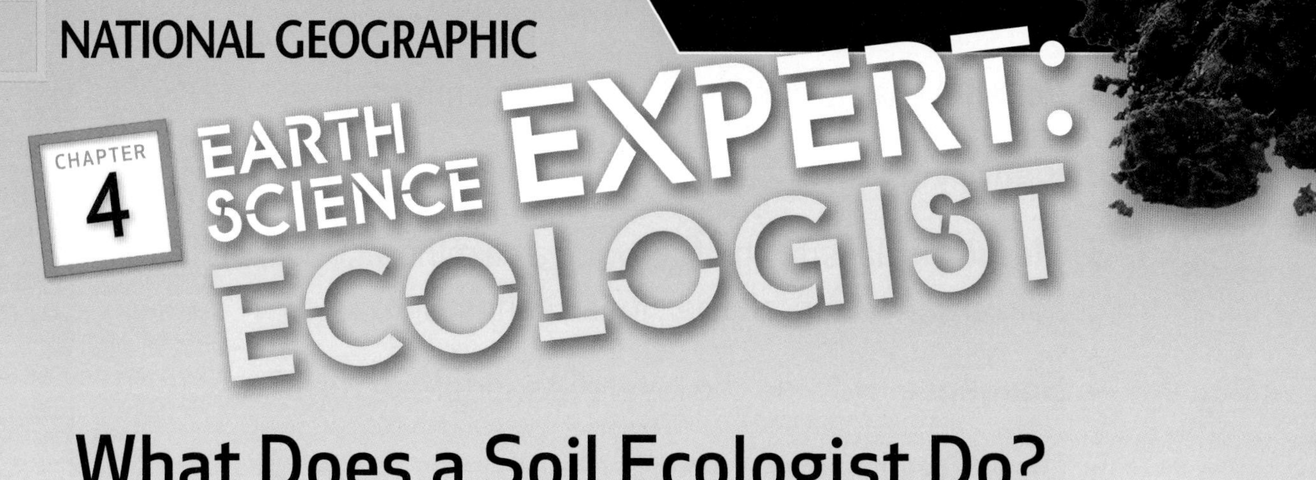

What Does a Soil Ecologist Do?

"Soil teems with life," says Louise Egerton-Warburton. It sure does. But most living things in soil are too tiny for you to see. It is a good thing Egerton-Warburton loves microscopes. She had one at an early age.

Egerton-Warburton studies things that live in the soil, such as fungi. A fungus can be large, like a mushroom. The fungi that Egerton-Warburton studies are too small to see. She collects them from outdoors. Sometimes she grows more in a lab.

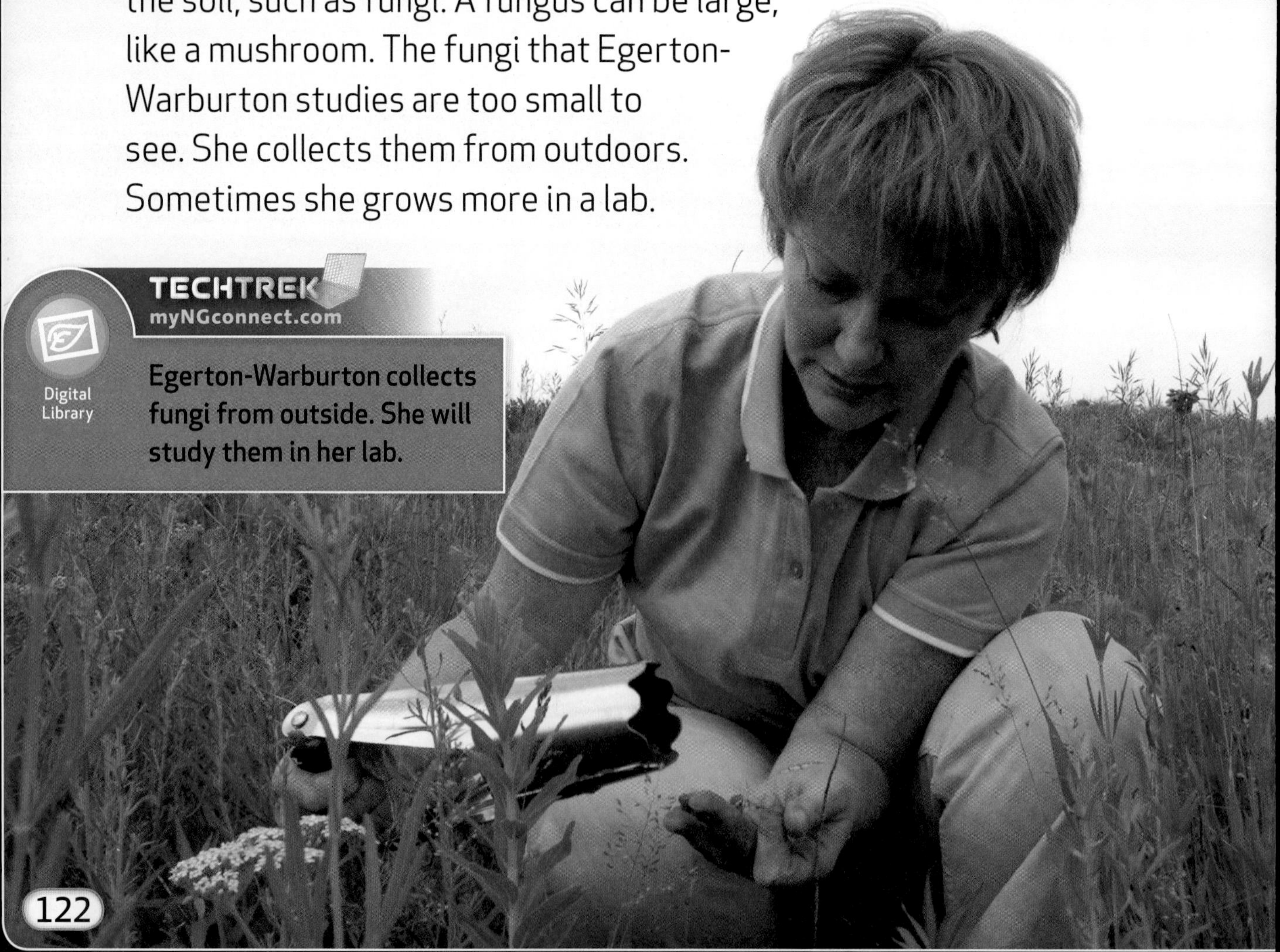

TECHTREK
myNGconnect.com
Digital Library

Egerton-Warburton collects fungi from outside. She will study them in her lab.

TECHTREK
myNGconnect.com
Student eEdition
Digital Library

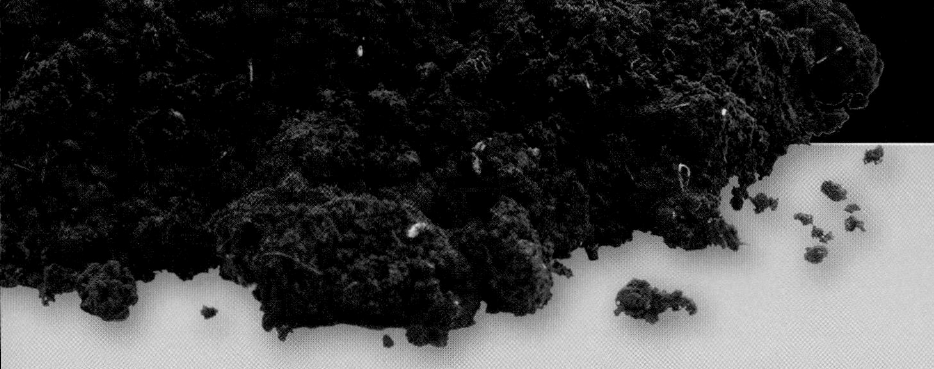

Then she studies them with a microscope. This tells her how they help plants get food and water. For example, fungi on roots can help two plants share water. Fungi also help plants get nutrients from the soil.

What is the coolest part of her job? She gets to use the most advanced technology. What is not so cool? Some living things smell really bad when they are growing. What is the most important part of her job? Conserving the soil. "Treating the soil kindly is the first step towards making sure we have food, clean water, and clean air." It also helps keep many different kinds of plants on Earth.

Would you like to be a soil ecologist? Egerton-Warburton studied science in school. Then she studied biology and chemistry in college. Her advice? "Do something that has always been fun and interesting."

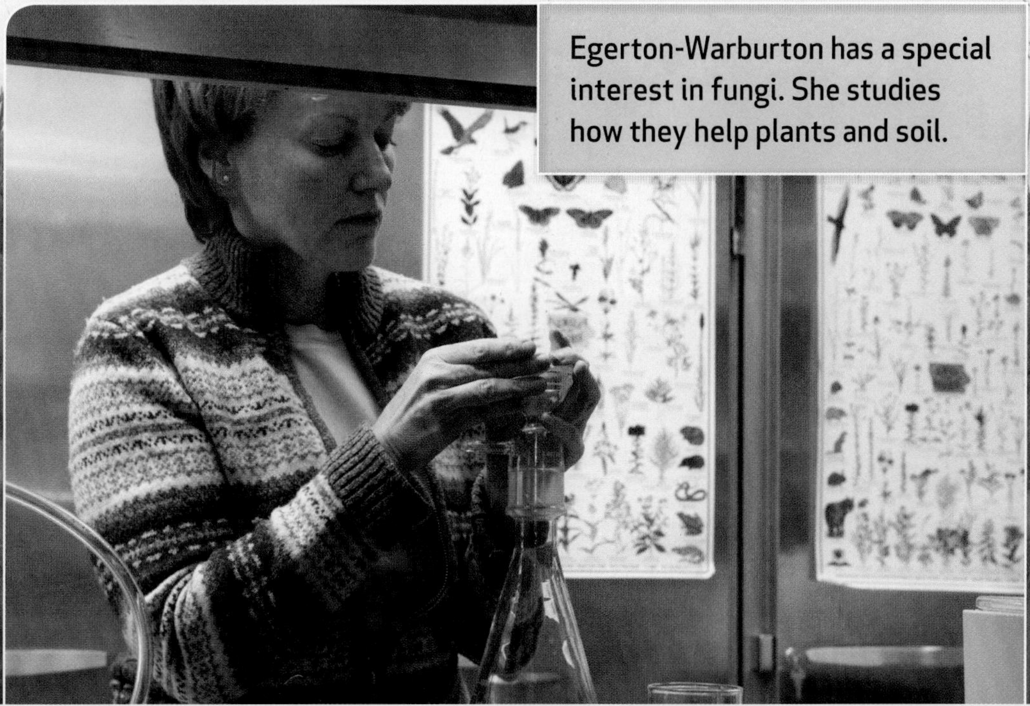

Egerton-Warburton has a special interest in fungi. She studies how they help plants and soil.

Famous Rocks: The World's Wonders

You might see rocks every day. But you probably haven't seen rocks like the ones on these pages. They are world-famous. Your guided tour begins here!

Uluru

The first stop is Australia to see Uluru (ūl-eh-RŪ), which is also called Ayers Rock. This is a huge rock formation in a desert. Uluru rises almost 350 meters (1,150 feet). That's taller than most skyscrapers. Suppose you stretched a rope around it. The rope would be 9 kilometers (5.6 miles) long.

Uluru is the Aboriginal name for Ayers Rock. Aboriginal people were the first to live here.

TECHTREK
myNGconnect.com
Student eEdition
Digital Library

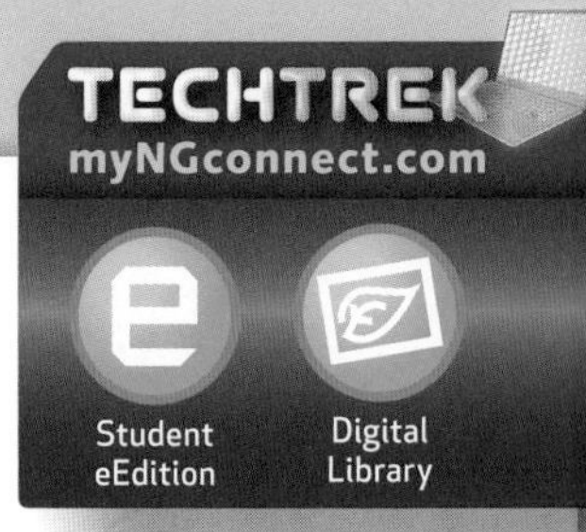

Uluru started as layers of sand. The sand built up and became thousands of meters deep. Over millions of years, the sand on the bottom layers hardened into sandstone. Uluru used to be part of a mountain range. Over time, water and wind wore away the mountains. All that remains is Uluru.

Like all sandstone, Uluru is made mostly of the **mineral** quartz. The large **grains** of this rock make it coarse.

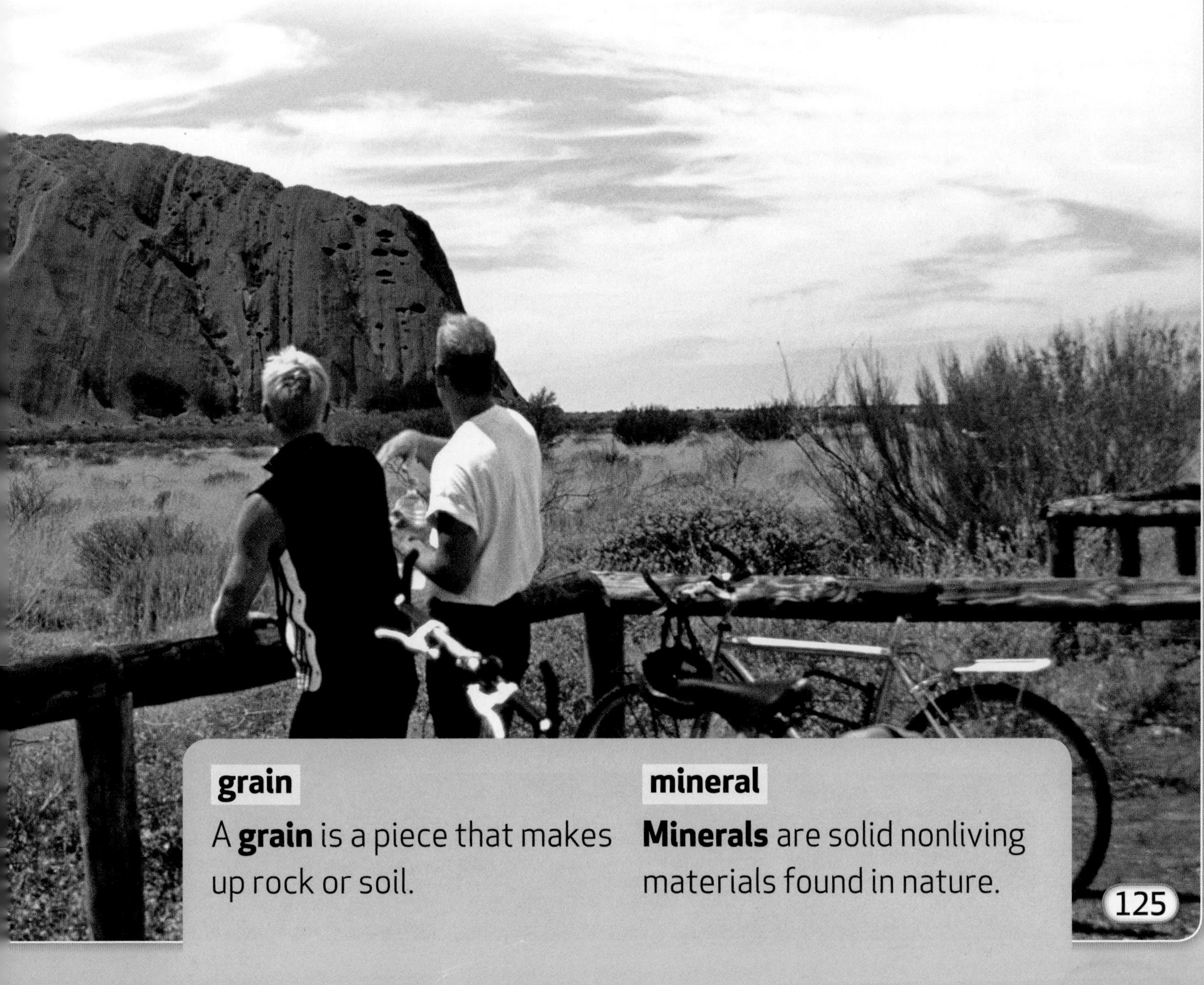

grain

A **grain** is a piece that makes up rock or soil.

mineral

Minerals are solid nonliving materials found in nature.

Giant's Causeway

Welcome to Giant's Causeway on the coast of Northern Ireland. It is a strange sight. Thousands of columns rise out of the sea onto the land. The columns are made of a rock called basalt. They look like tall pipes or bundles of thick rope. Most of the columns have six sides. An old Irish story says that the columns were built by giants. They wanted it as a road to Scotland. The word *causeway* means a "raised road."

These columns formed as rock cooled and shrank. The shrinking rock broke apart into a pattern. It's like the way soil cracks when it dries.

How did the Giant's Causeway really form? The six-sided rock columns were formed from lava, or melted rock, flowing from an ancient volcano. The lava cooled very quickly. As it cooled it formed the six-sided shapes we see today.

Tourists explore the columns of rock. About 40,000 columns make up the Giant's Causeway.

Karstic Peaks

Next, you visit Guilin, China. Karstic peaks tower over the city of Guilin. The peaks are made of limestone, a kind of rock. **Soil** covering the limestone has plenty of **humus** for plants to grow.

Like Uluru, these rocks formed in layers. But the rock did not form from sand or soil. Limestone comes from the shells of dead sea animals. This area was once under a sea. Over millions of years the shells built up. They stuck together and hardened into limestone.

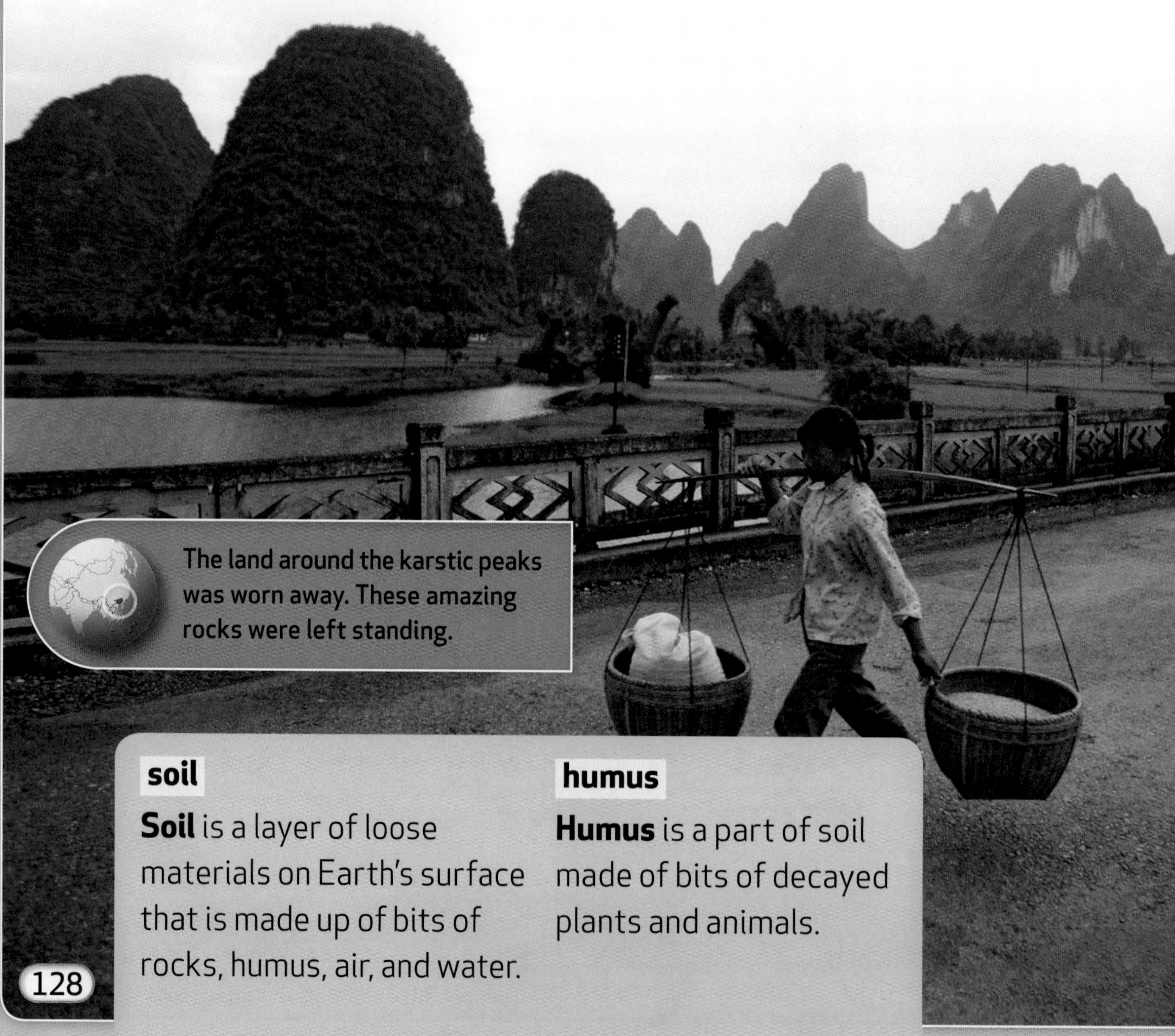

The land around the karstic peaks was worn away. These amazing rocks were left standing.

soil

Soil is a layer of loose materials on Earth's surface that is made up of bits of rocks, humus, air, and water.

humus

Humus is a part of soil made of bits of decayed plants and animals.

Devils Tower

Closer to home is Devils Tower in Wyoming. It looks like a giant tree stump. Like Giant's Causeway, Devils Tower is made of basalt.

Devils Tower formed near Earth's surface. Melted rock from deep inside Earth pushed up through cracks. The melted rock cooled and hardened. It was harder than the rock around it. Over thousands of years, the softer rock wore away.

Devils Tower is a national monument. The site is also special to Native Americans.

Sugar Loaf Mountain

Sugar Loaf Mountain is in Rio de Janeiro, Brazil. It is at the end of a range of mountains. Sugar Loaf Mountain rises 395 meters (1,296 feet) above the ocean. It's about as tall as the Empire State Building in New York City.

The mountain is made of granite rock. Granite forms from melted rock that hardens below ground. Forces deep inside Earth pushed the granite up to the surface. Then water and wind wore away the rock into its rounded shape.

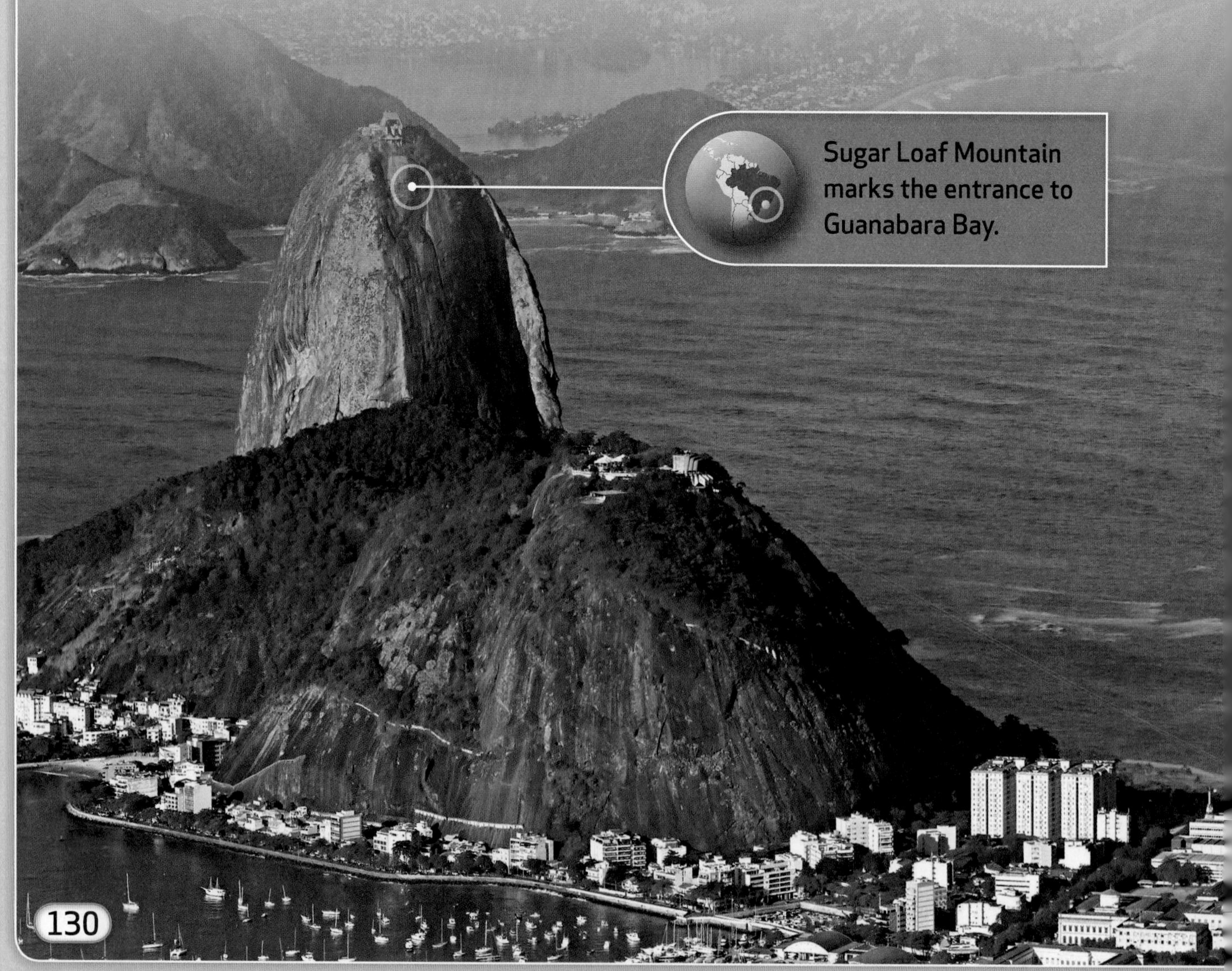

Sugar Loaf Mountain marks the entrance to Guanabara Bay.

Your tour doesn't have to end here. Amazing landforms are all over Earth. These are just a few. All of them are made of rock. Each took millions of years to form. But most formed in different ways. You can use the chart to compare them.

Digital Library

FAMOUS ROCKS AROUND THE WORLD

ROCK	TYPES OF ROCK	HOW FORMED
Uluru	sandstone	layers of sand pressed together
Giant's Causeway	basalt	lava on Earth's surface cooled and hardened
Karstic peaks	limestone	layers of shells stuck together
Devils Tower	basalt	melted rock cooled near Earth's surface
Sugar Loaf Mountain	granite	melted rock cooled below Earth's surface

NATIONAL GEOGRAPHIC

CHAPTER 4 SHARE AND COMPARE

Turn and Talk How are these famous rocks alike and different? Form a complete answer to this question together with a partner.

Read Select two pages in this section. Practice reading the pages. Then read them aloud to a partner. Talk about why the pages are interesting.

my SCIENCE notebook **Write** Write a conclusion that tells the important ideas about these famous rocks. Write the Big Idea in your own words. Share what you wrote with a classmate. Compare your conclusions.

my SCIENCE notebook **Draw** Form groups of five. Have each person draw one of the famous rocks and write a caption that describes how the rock formed. Combine your drawings and make a book about the famous rocks. Share your book with other groups in your class.

After reading Chapter 5, you will be able to:

- Recognize and describe different types of natural resources. **EARTH'S NATURAL RESOURCES**
- Identify and classify Earth's renewable resources. **RENEWABLE RESOURCES**
- Identify and classify Earth's nonrenewable resources. **NONRENEWABLE RESOURCES**
- Recognize that people depend on natural resources. **PEOPLE USE RESOURCES**
- Describe the harmful effects people have on the environment. **PEOPLE USE RESOURCES**
- Describe ways people are protecting, extending, and restoring natural resources. **PEOPLE CARE FOR RESOURCES**
- Science in a Snap! Describe ways people are protecting, extending, and restoring natural resources. **PEOPLE CARE FOR RESOURCES**

CHAPTER

5

WHAT ARE EART RES

You probably have a container for the supplies you need for school. Do you know there is a container for the supplies you need to live? Look at the picture. What do you think that container is? It's Earth! Earth has all the supplies or resources you need to live.

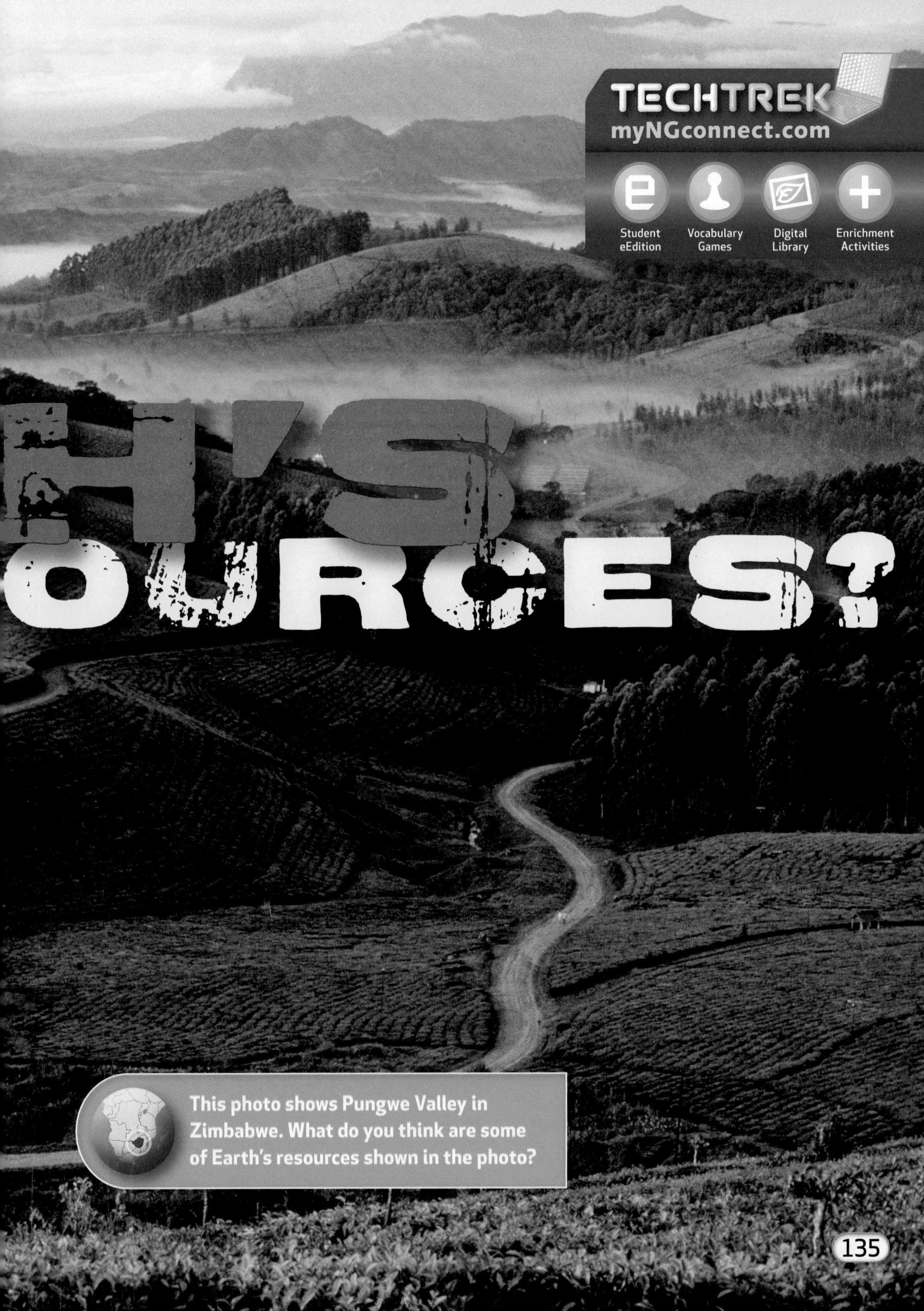

This photo shows Pungwe Valley in Zimbabwe. What do you think are some of Earth's resources shown in the photo?

SCIENCE VOCABULARY

natural resources
(NACH-ur-al RĒ-sors-es)

Natural resources are living and nonliving things found on Earth that people need. (p. 138)

Earth is rich in natural resources.

renewable resources
(rē-NŪ-ah-bl RĒ-sors-es)

Renewable resources are those that will not run out if used wisely. (p. 140)

Some materials, such as air, are renewable resources.

my Science Vocabulary

fossil fuel (FOS-il FYŪ-el)

natural resources (NACH-ur-al RĒ-sors-es)

recycling (rē-SI-kling)

nonrenewable resources (non-rē-NŪ-ah-bl RĒ-sors-es)

renewable resources (rē-NŪ-ah-bl RĒ-sors-es)

nonrenewable resources (non-rē-NŪ-ah-bl RĒ-sors-es)

Nonrenewable resources are those that cannot be replaced quickly enough to keep from running out. (p. 144)

Metals, such as copper, are nonrenewable resources.

fossil fuel (FOS-il FYŪ-el)

A **fossil fuel** is a source of energy formed from plants or animals that died long ago. (p. 146)

Coal is one fossil fuel that is burned for heat. The heat will be used to melt this metal.

recycling (rē-SI-kling)

Recycling is using the material in an old object to make a new object. (p. 154)

Paper is made from trees. We save trees by recycling used paper.

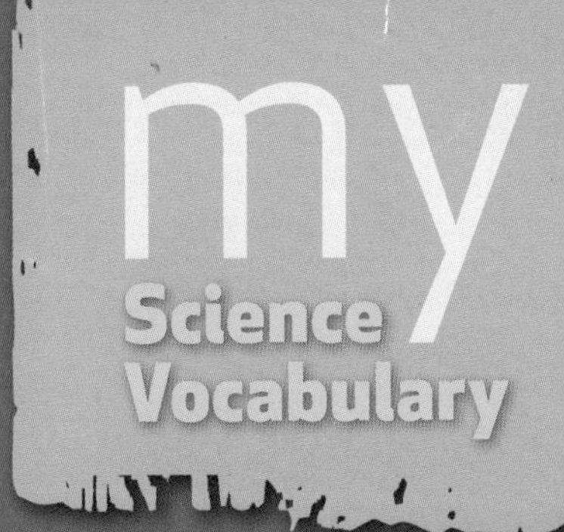

Earth's Natural Resources

Natural resources are living and nonliving things found on Earth that people need. Some resources are basic materials, such as air, water, and soil. Some are produced from basic resources, such as food, fuel, and building materials. Use the chart to learn more about natural resources.

People can find things they need at a market. Everything at the market comes from natural resources.

TECHTREK
myNGconnect.com
Digital Library

EXAMPLES OF **NATURAL RESOURCES**

FOSSIL FUEL

Fuel found in the ground heats many homes and runs cars. Fuels are a natural resource.

SOIL

Soil is a mixture of rocks, humus, air, and water. Most plants need soil to grow. Soil is a natural resource.

WATER

People need fresh water to drink. Water is a natural resource.

AIR

People need clean air to breathe. Air is a natural resource.

METALS

Metals are used to make things such as tools, building materials, and electronics. Metals are a natural resource.

Before You Move On

1. Define natural resources.
2. How is Earth similar to a supply box?
3. **Apply** Give three examples of ways you are using natural resources today at school.

Renewable Resources

Most people like sunny days. The sun provides light and energy to Earth. Energy is the ability to do work or cause a change. The sun's energy can change to heat. Plants use energy from the sun to make food for the plant.

The sun's energy will not run out anytime soon. It is a **renewable resource**. Renewable resources are those that will not run out if used wisely.

Hot air balloons move in the direction of the wind. Wind is also a renewable resource.

Air is an invisible mixture of gases. You can't see it, but you can see its effects. It can lift a huge balloon off the ground!

Air is another renewable resource. It is cleaned as it moves around Earth in winds. It is renewed when gases are used by plants and animals. Clean air is important to life on Earth.

A large burner heats the air inside the balloon.

Fresh water is another renewable resource. People and plants need water to live and grow. Look at the sprinklers in the photo. Why do you think they are needed? Rain does not always fall when the plants need it. Sometimes farmers need to water with sprinklers.

Sprinklers bring water to crops.

A forest is a renewable resource. Trees grow all the time to replace old ones. Seeds fall to the ground. Some of them grow into new trees. Like trees, other plants are renewable, too. Farmers can plant new crops each year.

People can help renew forests by planting new trees. The new trees replace those cut for wood.

Before You Move On

1. List four renewable resources.
2. Explain why sunlight is a renewable resource.
3. **Compare** Explain how forests renew themselves and how people renew forests. How are the results alike?

Nonrenewable Resources

Nonrenewable resources are those that cannot be replaced quickly enough to keep from running out. Metals and rocks are nonrenewable. Some kinds of rocks took millions of years to form. People remove the rocks from the ground. They use them to build things such as sidewalks and buildings. The supply of metals and rocks can get used up.

Look at the workers constructing a road. Concrete is pouring out of the truck. This thick mixture contains pieces of rock and other natural resources. The metal bars will help make the road stronger. The bars are made of a metal called iron. When the concrete hardens, cars will be able to drive over it.

Rocks like these are crushed and then used to make concrete for roads.

Coal, oil, and natural gas are fuels found on Earth. These are called **fossil fuels**. They are called that because they formed from fossils, or the remains of plants and animals that lived long ago. Fossil fuels are nonrenewable. These fuels power cars. They heat homes. People use them to make electricity.

These miners use heavy machinery inside a coal mine in Poland. They work deep underground to remove the coal.

People can also use the fossil fuel *oil* to make products. What are you wearing? If your clothes are made of nylon or polyester, they come from oil. Do you have plastic toys? Have you seen a plastic cell phone? Plastic is a material made from oil, too.

nylon backpack
plastic buckle
Which products in the picture come from nonrenewable resources?

Before You Move On

1. List five nonrenewable resources.
2. Explain why rocks are a nonrenewable resource.
3. **Draw Conclusions** Suppose there are no fossil fuels. Name at least three things in your classroom that would be missing. Explain why.

People Use Resources

People change the environment to meet their needs. Forests are cut down to make way for farm fields to grow food. Long ago this farm field may have been a forest. People cut down the trees to build homes, heat their homes, and make space for crops.

Today, farmers use machines, such as tractors, to help them grow crops on large areas of land.

People build new environments to meet their needs. Fields can be covered over by houses for people to live. Factories make products that people want. Shopping malls are built so people have a place to buy goods. Roads and bridges are built for transportation.

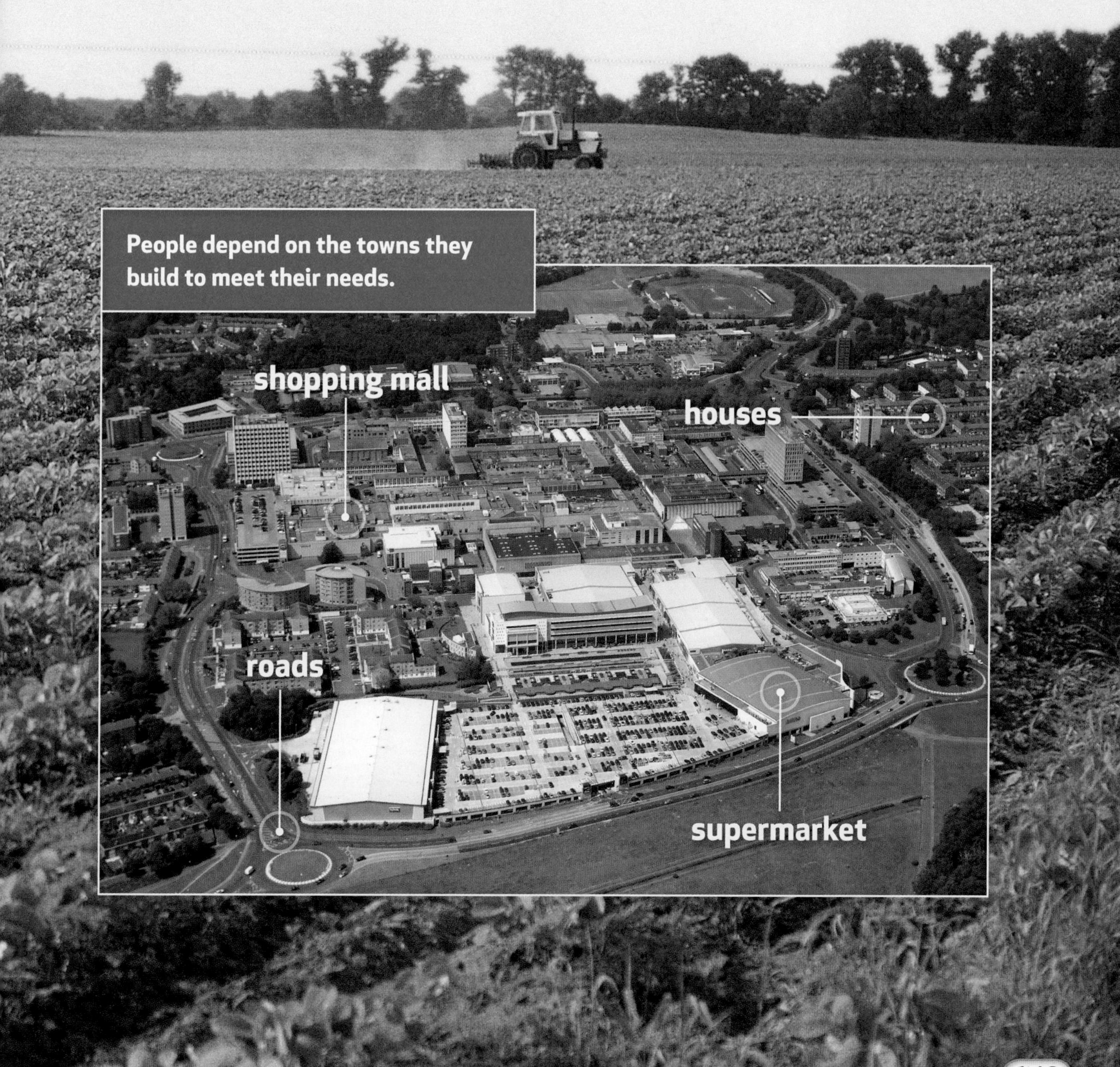

People depend on the towns they build to meet their needs.

How people use resources also can change the environment. For example, using Earth's resources can make trash. Plastic bags, boxes, and broken toys are collected and dumped in places called landfills. A landfill is a place where trash is buried. Landfills often use up large areas of land. They change the environment and destroy animal habitats.

In a landfill, plastic trash can last for about a million years. Glass bottles can last even longer.

Clearing land to build new homes changes the environment too. When a new neighborhood is built on the edge of a town, it can change the habitat. When the environment changes, many plants and animals cannot live there anymore. They must move away or else die out.

TECHTREK myNGconnect.com

Digital Library

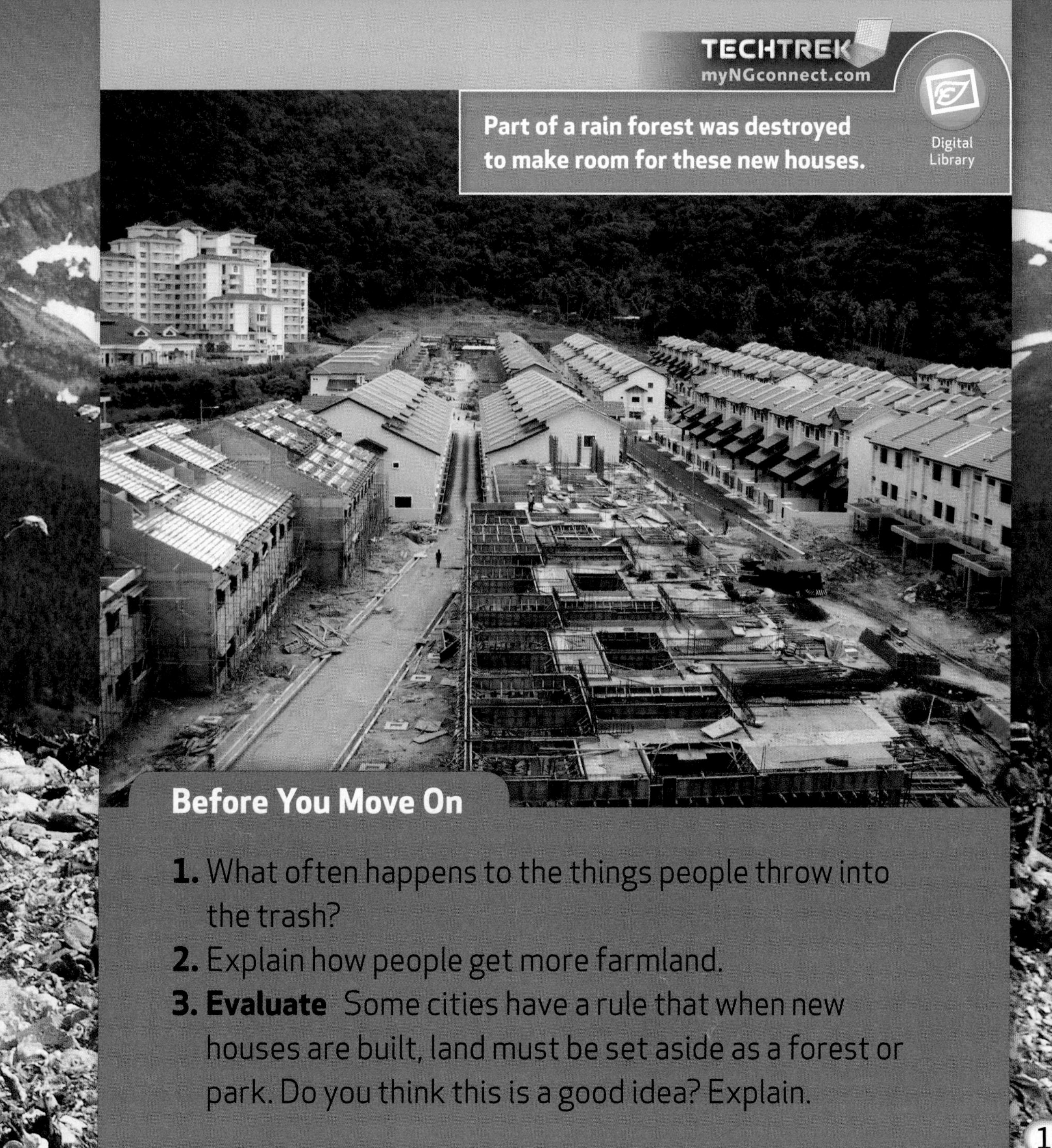

Part of a rain forest was destroyed to make room for these new houses.

Before You Move On

1. What often happens to the things people throw into the trash?
2. Explain how people get more farmland.
3. **Evaluate** Some cities have a rule that when new houses are built, land must be set aside as a forest or park. Do you think this is a good idea? Explain.

People Care for Resources

There are more people living on Earth than ever before. That is why it is more important to take good care of Earth's resources and not use them up. Many communities and industries practice land management. They are finding ways to use natural resources, such as rivers and forests, more carefully.

This is what the land in the large photo looked like when coal was mined there.

Sometimes damaged habitats can be repaired, or fixed. This is called land reclamation. Some holes made by mines can be filled in. New trees can be planted to replace those that were cut down. Then people can enjoy the area again. Plants and animals can return.

Compare this photo of reclaimed land in West Virginia with the photo on the left.

Do you throw old school papers in the trash? Do you throw away drink cans, jars, or magazines? These things can be recycled. **Recycling** means using the material in an old object to make a new object. For example, glass from an old jar can be used to make a new jar.

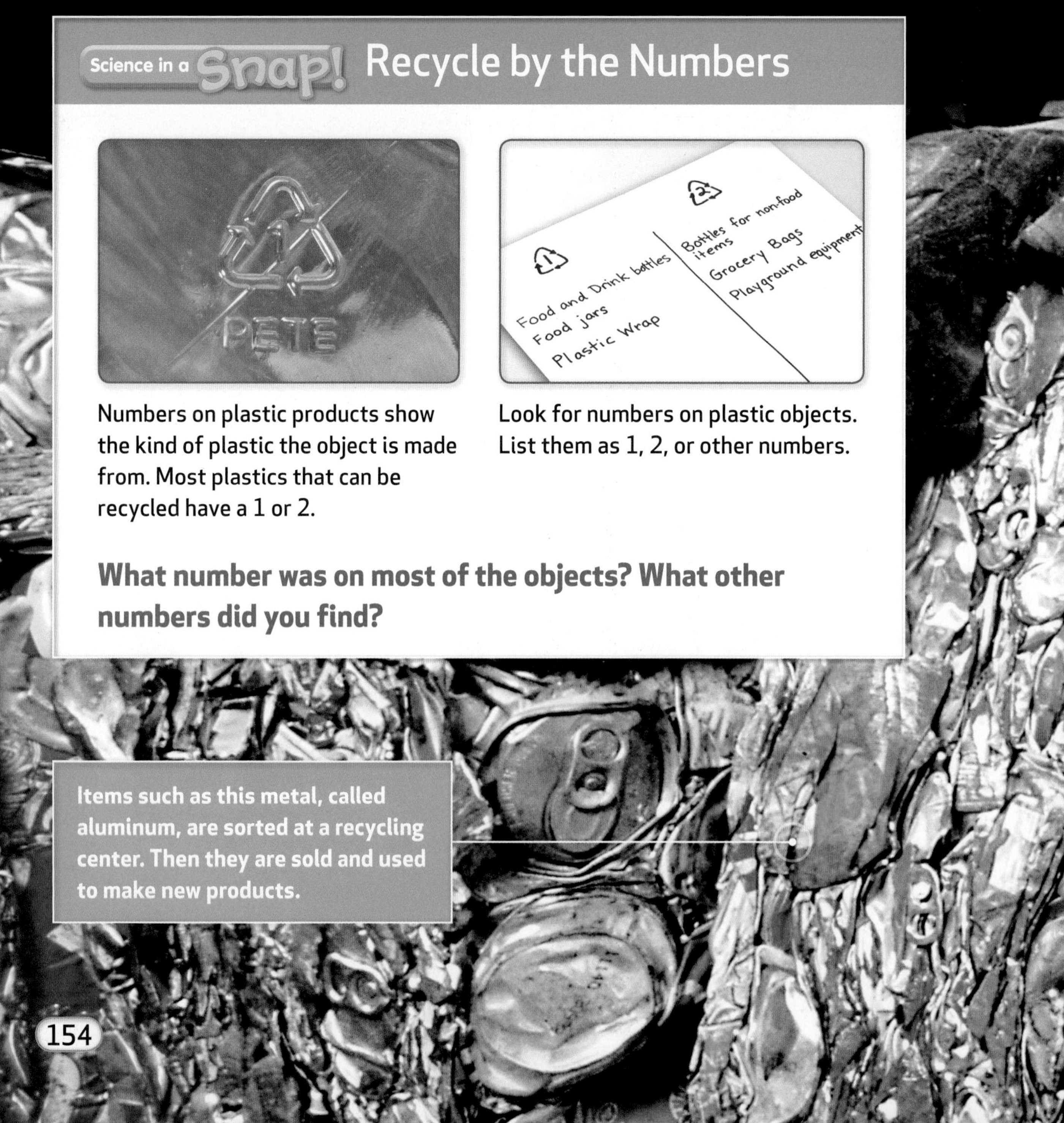

Science in a Snap! Recycle by the Numbers

Numbers on plastic products show the kind of plastic the object is made from. Most plastics that can be recycled have a 1 or 2.

Look for numbers on plastic objects. List them as 1, 2, or other numbers.

What number was on most of the objects? What other numbers did you find?

Items such as this metal, called aluminum, are sorted at a recycling center. Then they are sold and used to make new products.

TECHTREK
myNGconnect.com
Enrichment
Activities
WHAT CAN YOU RECYCLE?
PLASTIC
METAL
PAPER
GLASS

You can help care for Earth!

The things you do to save resources make a difference. But more people helping can make an even bigger difference! Talk to your friends and family. Work together to reduce, reuse, and recycle. Just use the three R's:

REDUCE

To use fewer resources, buy fewer things. Choose products with less packaging. Don't buy more than you need.

REUSE

Give away old toys and things that can be used again. Reuse shopping bags. Use worn-out clothes for rags.

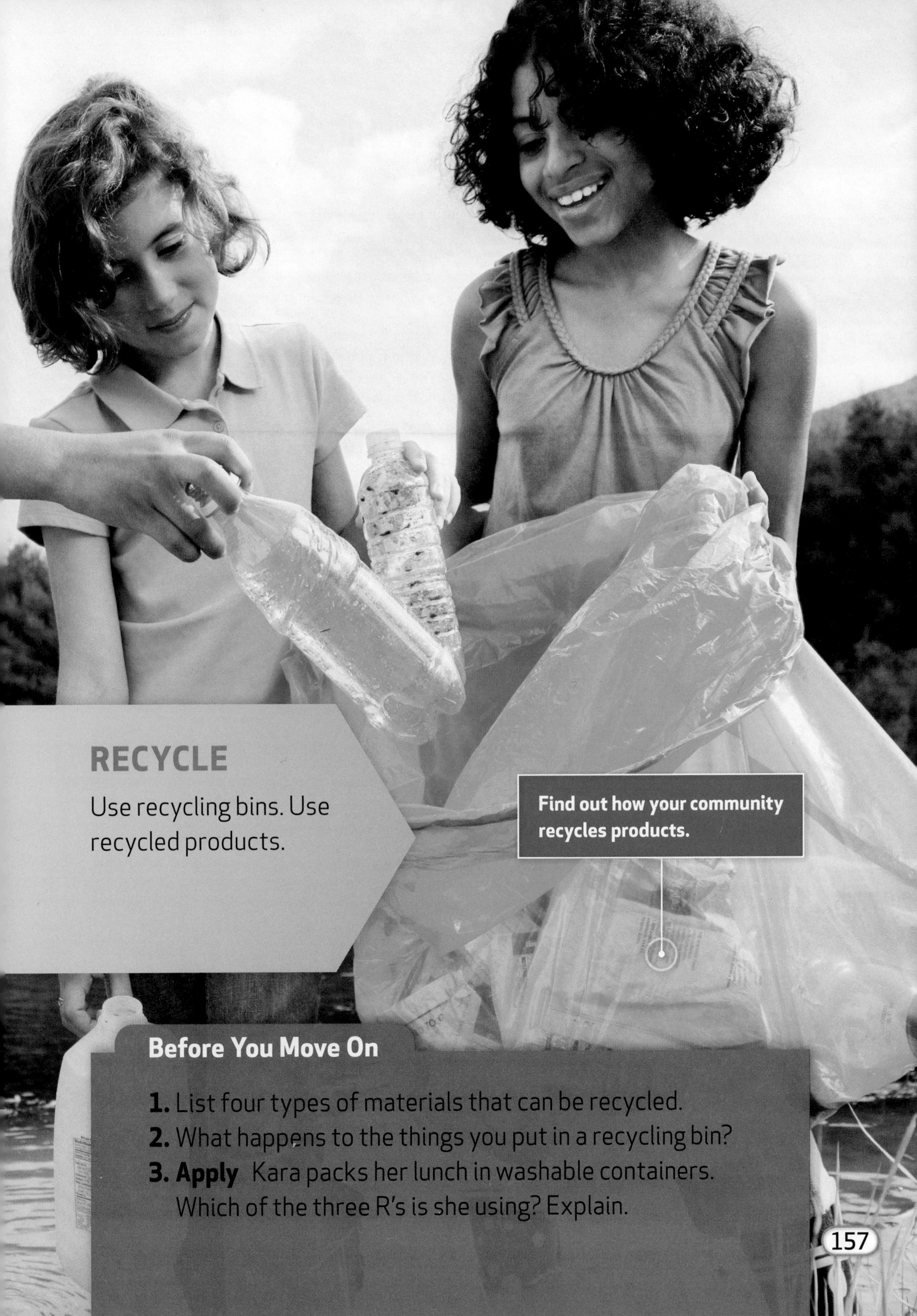

RECYCLE

Use recycling bins. Use recycled products.

Find out how your community recycles products.

Before You Move On

1. List four types of materials that can be recycled.
2. What happens to the things you put in a recycling bin?
3. **Apply** Kara packs her lunch in washable containers. Which of the three R's is she using? Explain.

NEW YORK CITY: NOW AND THEN

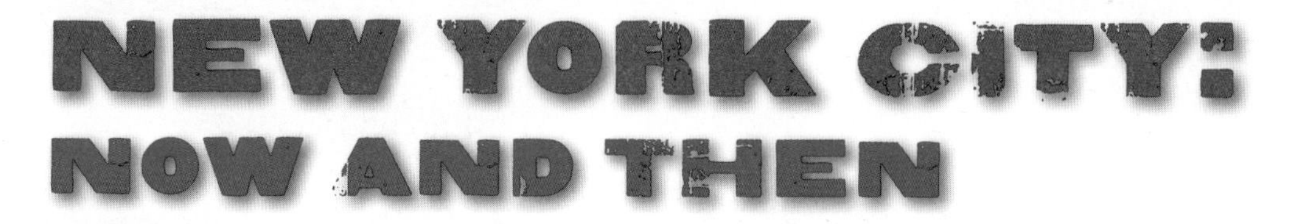

Have you ever been to New York City? People go there to see the Statue of Liberty. They see shows on Broadway. Lights shine, cars honk, and people are everywhere. New York City is the biggest city in the United States. Below, you can see the part of New York City that is located on Manhattan Island.

New York City is on the Atlantic coast at the mouth of the Hudson River. More than eight million people live there.

Now turn back the clock to the time before New York City was built. Manhattan Island didn't look like it does today. Forests covered the land. Streams ran through the forests. Elk, black bears, and turkeys lived there. It was quiet and peaceful. The picture shows how this area might have looked before people changed the land to build New York City.

Much of this land was once a salt marsh habitat.

Conclusion

Earth has all the resources people need to live. It has sunlight, plants, fossil fuels, and more. Some resources are renewable. Others are nonrenewable. People must care for Earth's resources so they don't run out. People can help by using the three R's—reduce, reuse, and recycle.

Big Idea Earth's resources are the materials found on Earth that people need and use, such as sunlight, water, plants, soils, rocks, metals, and fuel.

Renewable Resources

+

Nonrenewable Resources

= EARTH'S RESOURCES

Vocabulary Review

Match the following terms with the correct definition.

A. recycling
B. natural resources
C. nonrenewable resources
D. fossil fuel
E. renewable resources

1. Living and nonliving things found on Earth that people need
2. Using the material in an old object to make a new object
3. Resources that cannot be replaced quickly enough to keep from running out
4. Resources that will not run out if used wisely
5. A source of energy formed from plants or animals that died long ago

Big Idea Review

1. **Recall** What resource is plastic made from? Is the resource renewable or nonrenewable?

2. **Describe** What are three things you could do to make less trash?

3. **Contrast** What is the difference between renewable and nonrenewable resources?

4. **Classify** Tell whether each object was made from a renewable or a nonrenewable resource: *cotton socks, a glass bottle, notebook paper, a silver necklace.*

5. **Apply** Imagine you are in charge of a large forest. Every year, some of your trees are cut down to make paper. What can you do to make sure the trees will never run out?

6. **Analyze** Suppose an old gas station was torn down near your home. What would you do to the land to make it into a park?

Write About Forest Resources

Explain How do you think people have changed this area of forest? Can the resource renew itself? How? Can people help renew it? How?

Want to Make a Difference? Teach About the Environment!

Do you want to make Earth healthier? So does Gretchen Gigley. She works for a group of people who want cleaner air.

What do you do in your career?

I go to schools and talk to parents, teachers, and students. I teach them how to help clean the air. Clean air makes people, plants, animals, and the land healthier.

What motivates you the most?

This career allows me to help to make a difference in the environment and in my community. The lessons learned by students in our programs encourage them to be friendlier to the environment. By taking public transportation, and walking and biking, young people can make a big difference in reducing air pollution in their communities.

Gretchen Gigley uses her science background to teach about the environment.

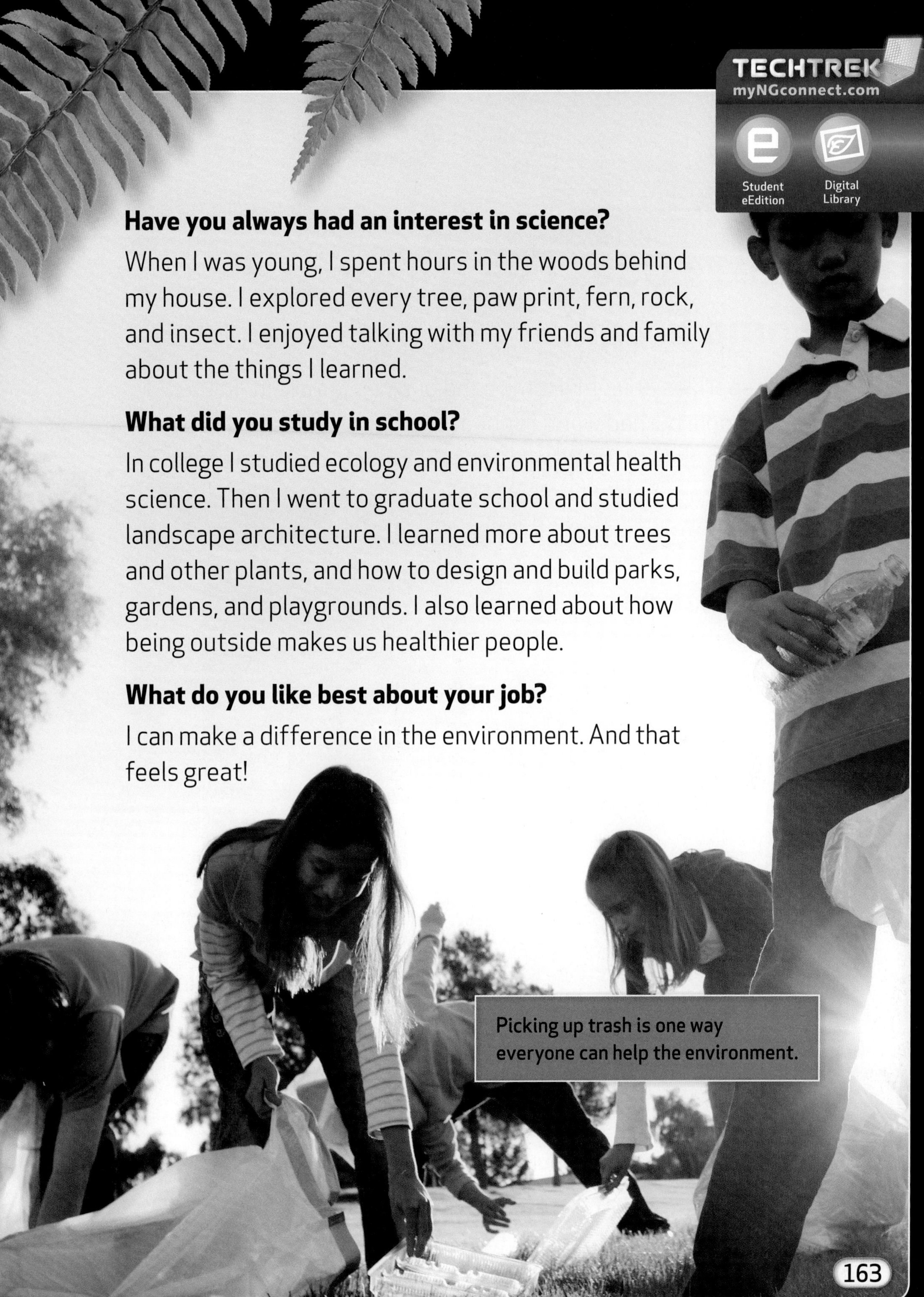

Have you always had an interest in science?

When I was young, I spent hours in the woods behind my house. I explored every tree, paw print, fern, rock, and insect. I enjoyed talking with my friends and family about the things I learned.

What did you study in school?

In college I studied ecology and environmental health science. Then I went to graduate school and studied landscape architecture. I learned more about trees and other plants, and how to design and build parks, gardens, and playgrounds. I also learned about how being outside makes us healthier people.

What do you like best about your job?

I can make a difference in the environment. And that feels great!

Picking up trash is one way everyone can help the environment.

NATIONAL GEOGRAPHIC

BECOME AN EXPERT

Where Does Electricity Come From?

Earth sparkles at night. Homes and businesses are lit. In the past, people burned wood for heat and light. Today most people depend on electricity. Where does that electricity come from? It comes from **natural resources**.

This picture of North America was taken from space. It shows the lights from homes and businesses at night.

natural resources

Natural resources are living and nonliving things found on Earth that people need.

TECHTREK
myNGconnect.com
Student eEdition
Digital Library

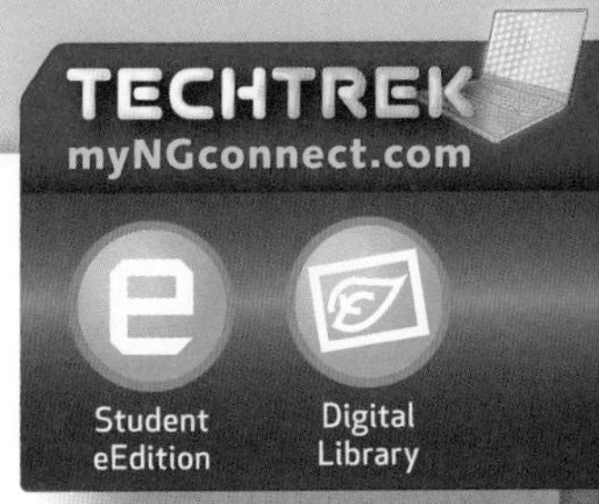

Electricity is used for everything from drying your hair to **recycling** material in this factory. Common objects you use every day, such as bottles, cans, and paper, are made with electricity.

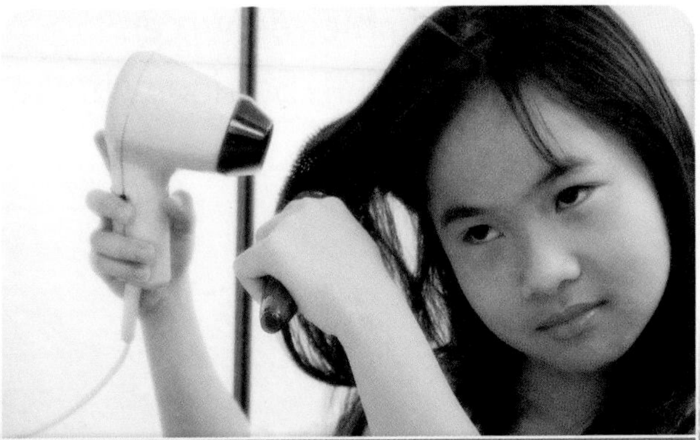

This girl uses electricity to make her hair dryer work.

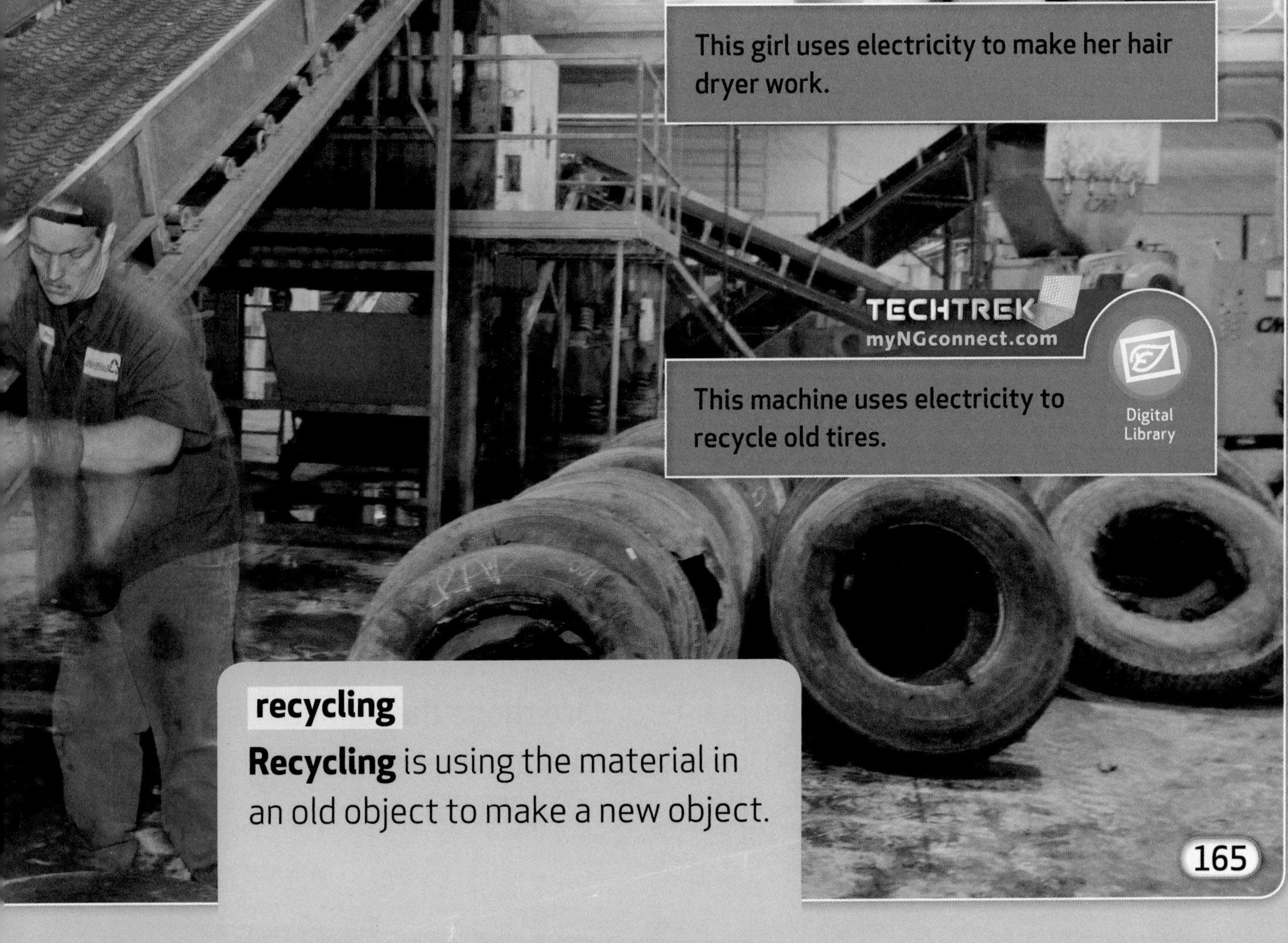

TECHTREK
myNGconnect.com
Digital Library

This machine uses electricity to recycle old tires.

recycling

Recycling is using the material in an old object to make a new object.

Fossil Fuels

Most electricity today comes from **fossil fuels**. Coal is used most often. Coal is found in the ground. People dig mines and remove the coal. Coal is a **nonrenewable resource**. We use a lot of it. Someday it will run out.

This coal will be burned to help make electricity.

COAL TO ELECTRICITY

1. Coal is burned to heat water and make steam.

2. The steam runs machines that make electricity.

3. The electricity goes to you.

fossil fuel

A **fossil fuel** is a source of energy formed from plants or animals that died long ago.

nonrenewable resources

Nonrenewable resources are those that cannot be replaced quickly enough to keep from running out.

Nuclear Power

Some electricity comes from uranium or other nuclear fuels. Like coal, nuclear fuels are found in the ground. They are a nonrenewable resource. Nuclear power does not make the air dirty. Look at the photo below. The clouds coming from the power plant are made of steam, not smoke. When nuclear fuel is used, waste materials are left over. Nuclear waste is poisonous to people. It must be stored safely.

A nuclear power plant, such as this one, uses nuclear fuel to heat water and make steam. Then the steam runs machines that make electricity.

Solar Power

The sun's energy is a **renewable resource**. The sun provides light to Earth. When sunlight strikes solar panels, the sun's energy changes to electricity. Some solar power plants use the sun's energy to heat water and make steam. Then the steam runs machines that make electricity. Many solar panels are needed to power towns and cities. The solar panels can take up a lot of land.

TECHTREK myNGconnect.com Digital Library

Solar panels collect energy from the sun.

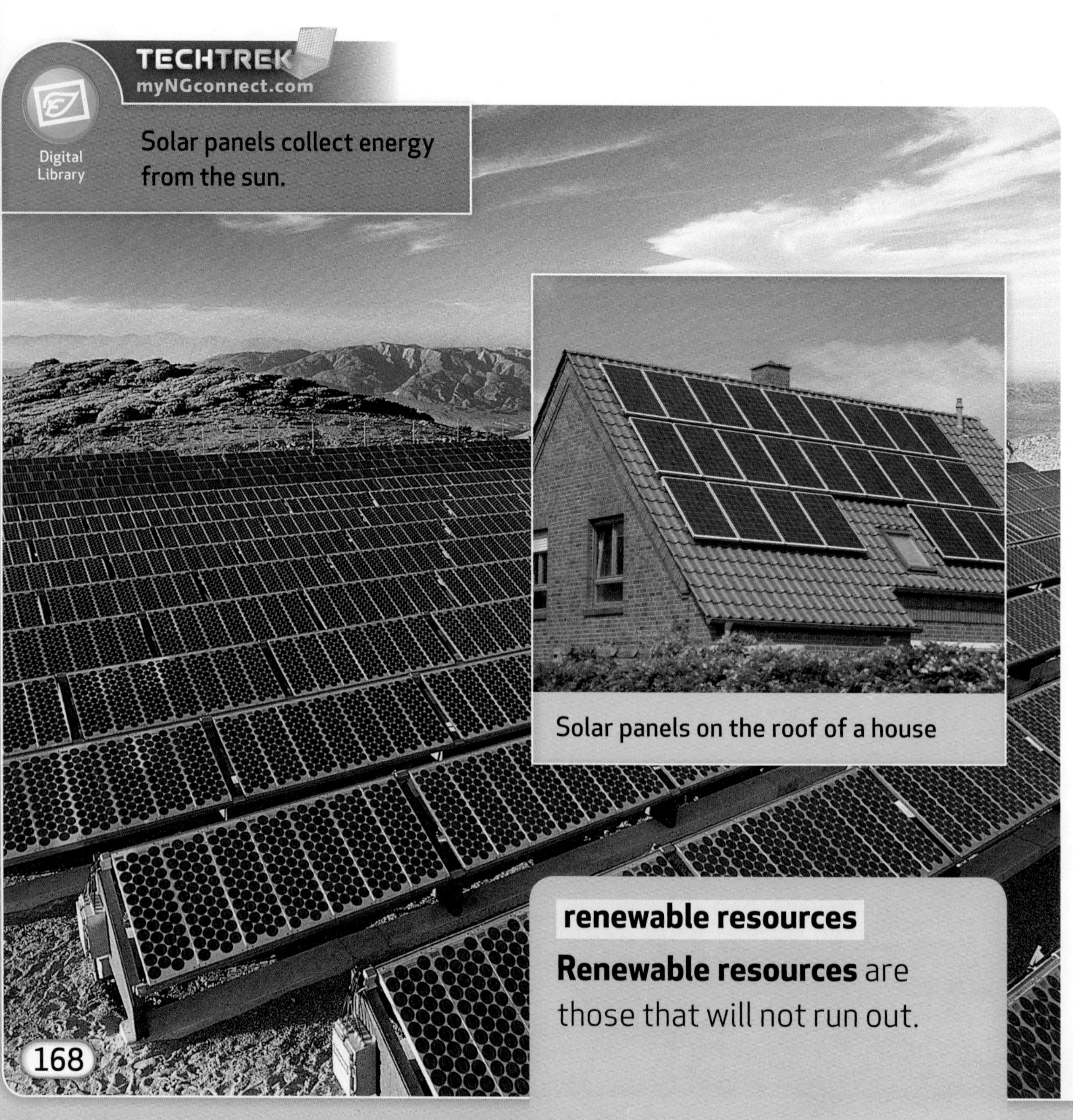

Solar panels on the roof of a house

renewable resources

Renewable resources are those that will not run out.

Wind Power

Wind can be used to make electricity. Wind is renewable. It will not run out. The wind's energy is used to turn the blades of a machine called a turbine. As the wind blows, it pushes on the turbine's blades. As the blades turn, they run a machine that makes electricity. Wind turbines work best in places where winds are strong. But they can be noisy. Some people think wind turbines are ugly.

WIND TO ELECTRICITY

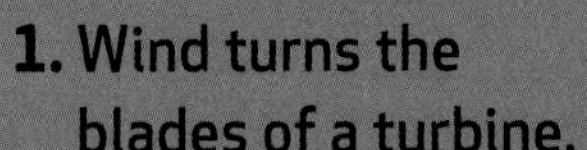
1. Wind turns the blades of a turbine.

2. The turbine runs a machine that makes electricity.

3. The electricity goes to you.

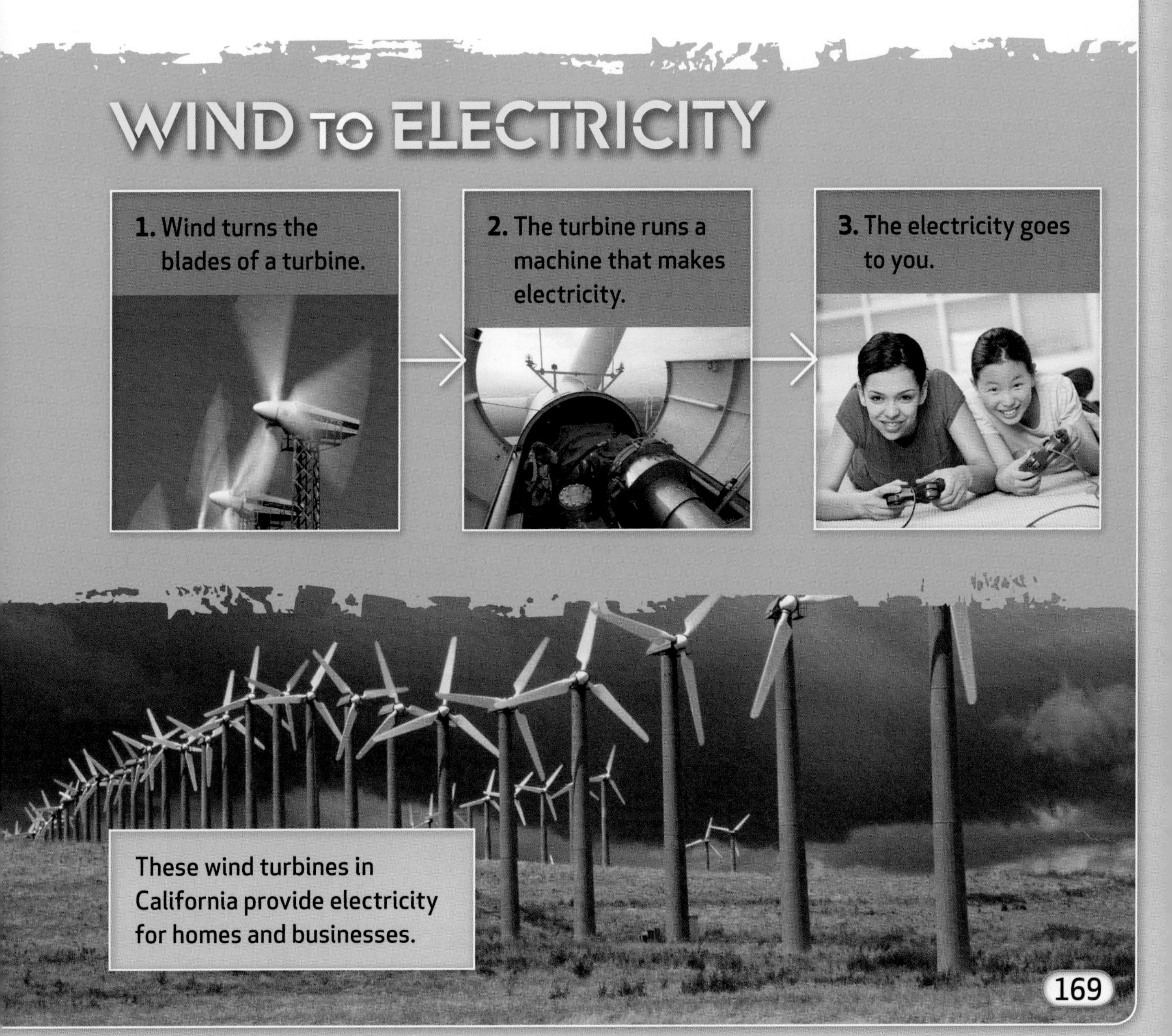

These wind turbines in California provide electricity for homes and businesses.

Hydroelectric Power

"Hydroelectricity" means water electricity, because "hydro" means water. We can use the water that rushes over a dam to make electricity. The moving water pushes on the blades of turbines inside the dam. As the blades turn, they run machines that make electricity.

This dam was built on the Payette River near Cascade, Idaho.

People use different natural resources to make electricity. Some are renewable, such as the sun, wind, and water. Others are nonrenewable, such as coal and uranium. People depend on these resources to make the electricity they use every day.

WATER TO ELECTRICITY

1. Water rushes over a dam.

2. The rushing water runs the machines that make electricity.

3. The electricity goes to you.

NATIONAL GEOGRAPHIC

SHARE AND COMPARE

Turn and Talk Where does electricity come from? Form a complete answer to these questions together with a partner.

Read Select two pages in this section. Practice reading the pages. Then read them aloud to a partner. Talk about why the pages are interesting.

my SCIENCE notebook **Write** Write a conclusion that tells the important ideas about where electricity comes from. State what you think is the Big Idea of this section. Share what you wrote with a classmate. Compare your conclusions.

my SCIENCE notebook **Draw** Draw a picture of yourself doing something at home that uses electricity. Label your drawing. Then write how much time you think you spend doing it. Combine your drawing with those of your classmates to imagine how much people use electricity.

After reading Chapter 6, you will be able to:

- Recognize that weathering and erosion are processes that change Earth's surface slowly. **WEATHERING, EROSION AND DEPOSITION**
- Recognize that wind, water, ice and plants weather rock. **WEATHERING**
- Recognize that wind, water, and ice cause erosion. **EROSION AND DEPOSITION**
- Describe how eroded material can be deposited in a new place. **EROSION AND DEPOSITION**
- Identify some of Earth's landforms. **LANDFORMS ON EARTH'S SURFACE**
- Science in a Snap! Recognize that wind, water, and ice cause erosion. **EROSION AND DEPOSITION**

CHAPTER

6

HOW DO WEATHERING AND EROSION CHANGE

Thousands and thousands of years ago, the rocks in the picture were all connected. They were part of one huge rock. Over the years, water and wind slowly wore away much of the rock. The remaining rocky towers still change a little bit each day. Can you tell how?

THE LAND?

Wind blows sand across the desert in western Australia. How do you think the blowing sand changes the land?

SCIENCE VOCABULARY

weathering (WETH-urh-ing)

Weathering is the breaking apart, wearing away, or dissolving of rocks. (p. 178)

Over time, weathering caused the rocks to get smaller.

erosion (ē-ROH-zhun)

Erosion is the picking up and moving of rocks and soil to a new place. (p. 182)

Ocean waves cause erosion along this coastline.

glacier (GLĀ-shur)

A **glacier** is a huge area of ice that moves slowly over Earth's surface. (p. 184)

This glacier picks up rocks and soil as it moves.

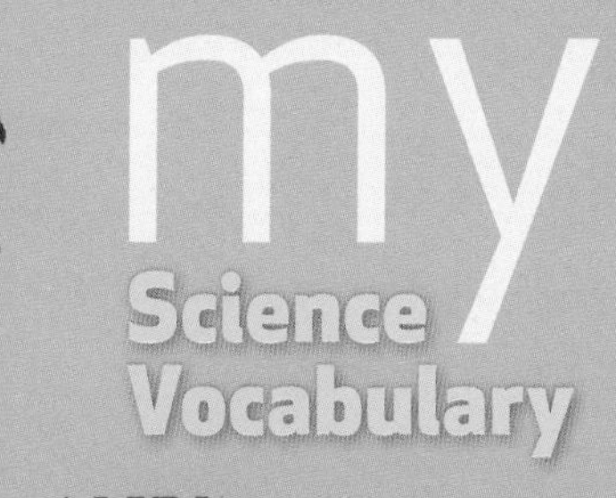

deposition
(dep-uh-ZISH-uhn)

erosion
(ē-ROH-zhun)

glacier
(GLĀ-shur)

landform
(LAND-form)

weathering
(WETH-urh-ing)

deposition
(dep-uh-ZISH-uhn)

Deposition is the laying down of rock and soil in a new place. (p. 185)

Deposition happened when a glacier melted and left these rocks behind.

landform (LAND-form)

A **landform** is a natural feature on Earth's surface. (p. 186)

A hill is one kind of landform.

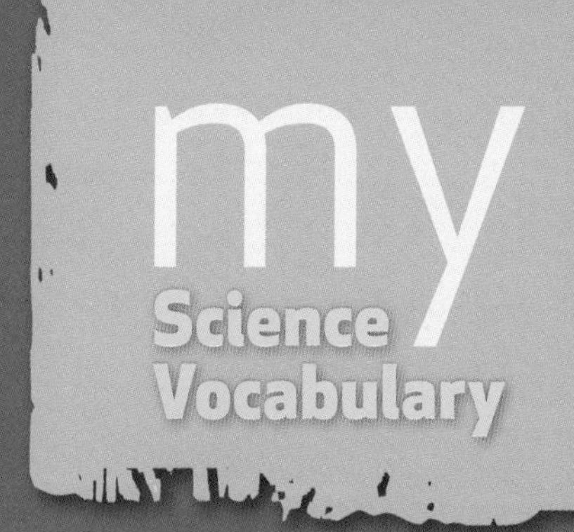

Weathering

Wind, plants, water, and ice can cause rocks to break apart, wear away, or dissolve. This breaking apart, wearing away, or dissolving of rock is called **weathering**.

Wind, for example, blows sand against rocks. The sand chips away at the rock, breaking off little pieces.

Weathering by wind helped to shape this rock. What do you think it looks like?

Plants can cause weathering, too. Plants sometimes grow in the cracks of rocks. The plant roots grow wider and deeper. They push against the sides of the crack. Over time, the crack becomes bigger and bigger until the rock splits.

The growing roots and trunk of this tree are breaking the rock apart.

Water is another cause of weathering. Raindrops pound at rock, wearing it away bit by bit. Streams and rivers pound at rocks, too. The rushing waters carry pebbles that scrape against rocks on the river bottom. Over time, this weathering can break large rocks into smaller and smaller rocks.

Water rushes over rocks and wears them away in a fast-moving stream.

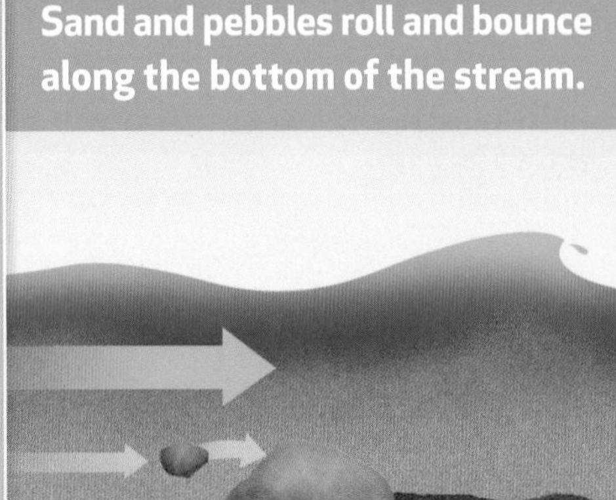

Sand and pebbles roll and bounce along the bottom of the stream.

The sand and pebbles strike other rocks and break them.

Lakes and oceans play a part in weathering. Waves slam against rocks and break off small pieces. Frozen water does its share of weathering, too. When water freezes inside cracks in rocks, the ice pushes against the sides of the cracks. Over many years of freezing and thawing, the rock splits apart.

Weathering by ice loosened rocks on Devils Tower in Wyoming. The loose rocks fell to the base of the tower.

Before You Move On

1. How does weathering change Earth's surface?
2. How are weathering by wind and weathering by plants alike and different?
3. **Infer** What do you think will happen to the shape of Devils Tower over a long period of time?

Erosion and Deposition

Erosion Pieces of weathered rock usually don't stay in one place. Wind, water, and ice carry them to a new place. **Erosion** is the picking up and moving of rocks and soil. Wind causes erosion by blowing soil and sand around. The picture shows how running water causes erosion. Streams and rivers carry away loose, weathered material. This action carves out valleys big and small.

This stream in Kansas carves out a small valley as the water carries rocks and soil away.

Ocean water erodes a lot of material. After waves break off pieces of rock, the water carries it to different places along the shore. How are waves eroding the seashore below?

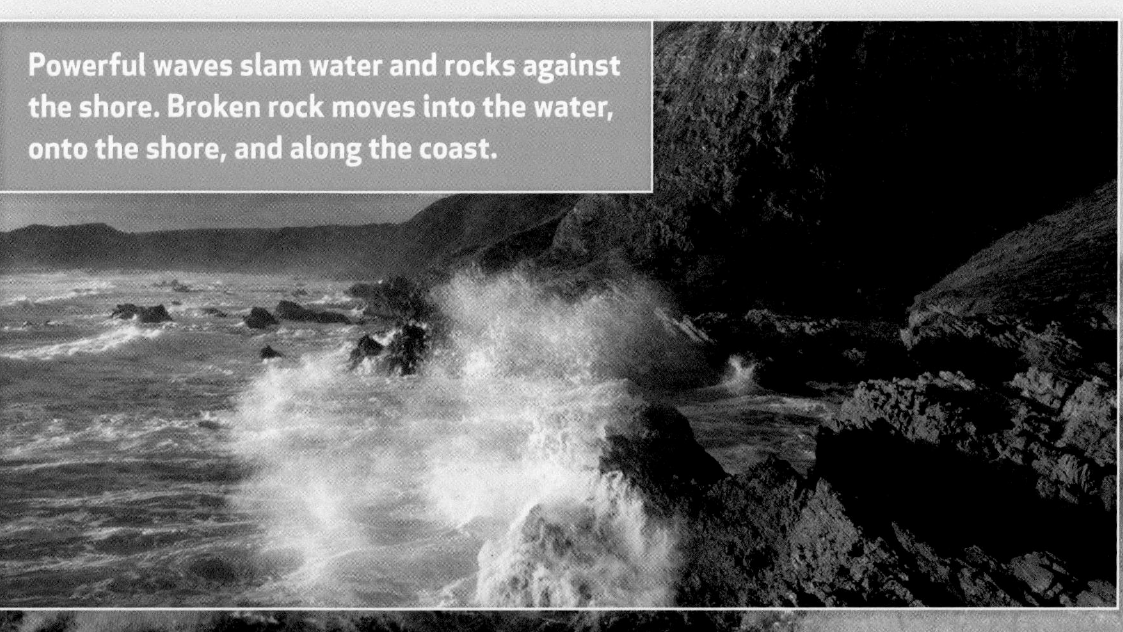

Powerful waves slam water and rocks against the shore. Broken rock moves into the water, onto the shore, and along the coast.

Science in a Snap! Along the Shore

Pour half a cup of sand at one end of a container. Pour half a cup of water at the other end.

Gently move the container back and forth so that water sloshes against the sand. Observe how the sand moves.

What do you think happens to sand along a beach? Use this model to help you answer.

Ice also does the work of erosion. A **glacier** is a huge area of ice that moves slowly over Earth's surface. Glaciers move downhill, just as rivers do. As the glaciers move, they pick up large amounts of soil and rocks. The glaciers carry these materials as the ice moves.

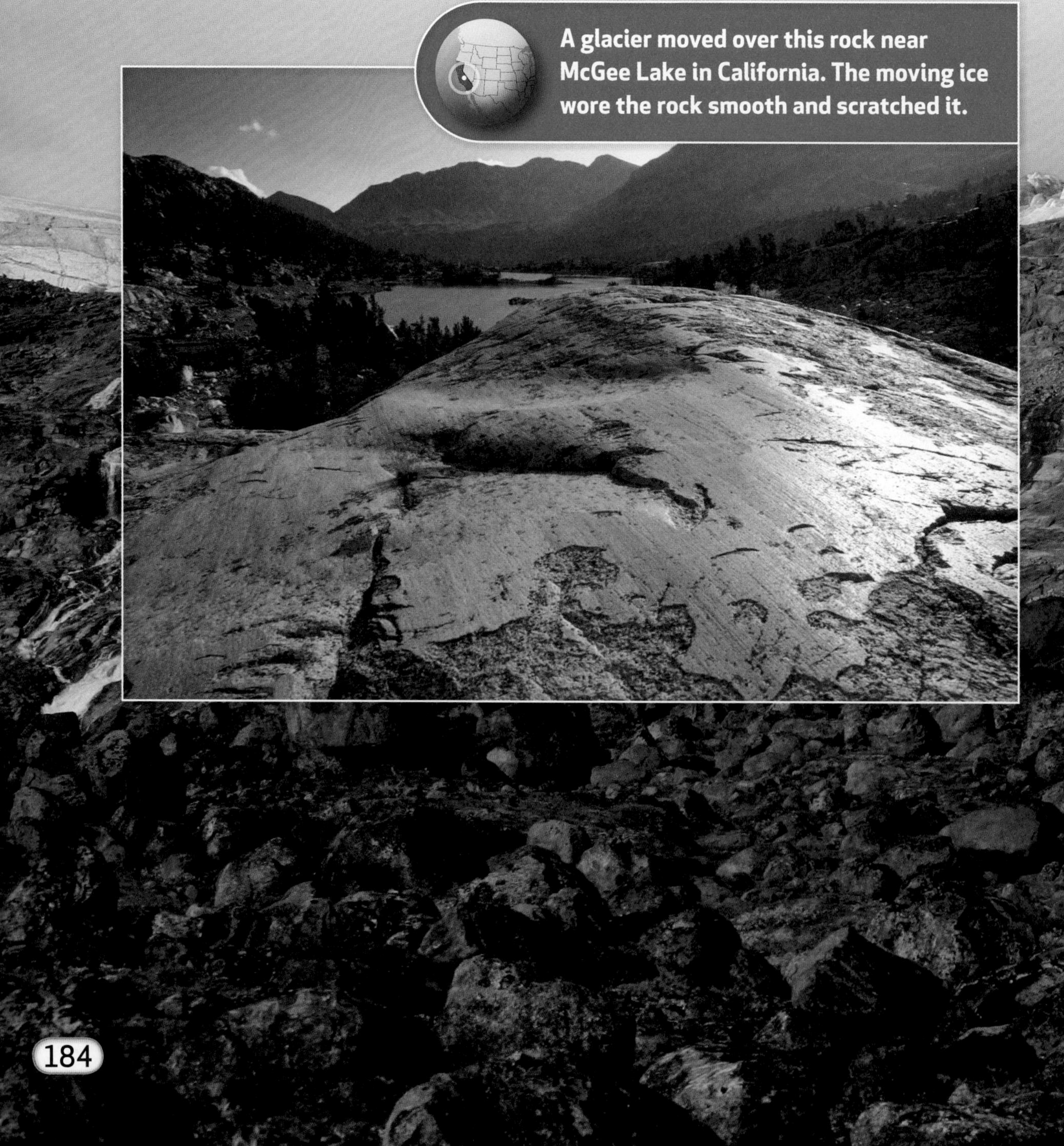

A glacier moved over this rock near McGee Lake in California. The moving ice wore the rock smooth and scratched it.

Deposition When glaciers melt, the material they were carrying drops to the ground. This is called **deposition**, or the laying down of rocks and soil in a new place. Deposition also happens when a river slows down as it enters a lake or ocean. The river drops much of its mud and rocks. New land might build up. Deposition along a shore makes a beach. Wind deposition makes sand dunes.

These rocks were carried in the glacier you see in the background. As the glacier melted, it deposited the rocks in this place.

Before You Move On

1. What is erosion?
2. Give an example of deposition.
3. **Predict** If you pour water into the cracks of rock in cold weather, what do you think will happen?

Landforms on Earth's Surface

Weathering, erosion, and deposition slowly shape and reshape the land over many years. These processes help make the land look the way it does. In other words, they shape **landforms**. A landform is a natural feature on Earth's surface.

When you travel in the United States, you can see many landforms. You can ride over hills and mountains. You can cross plains and valleys. You will see canyons and plateaus. These are all landforms.

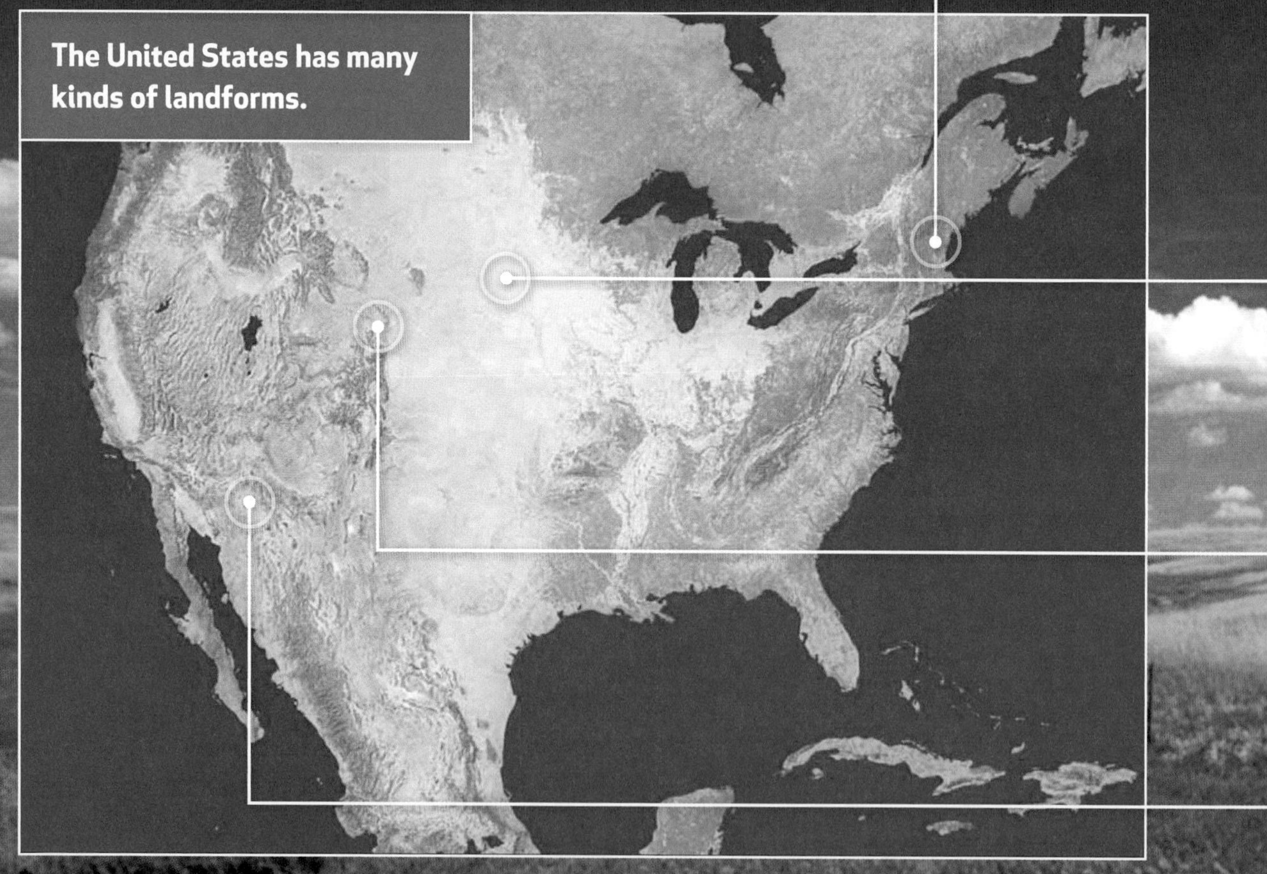

The United States has many kinds of landforms.

Digital Library

HILL

A hill is a raised mound of land smaller than a mountain. These hills are in New Hampshire.

PLAIN

A plain is a large flat area of land. These plains are part of the Fort Pierre National Grassland in South Dakota.

MOUNTAIN

A mountain is a very high place with steep sides that rises above the surrounding land. These mountain are the Tetons in Wyoming.

PLATEAU

A plateau is a raised area of flat land. This plateau is in Arizona.

Landforms can be very different from each other, even if they are the same kind of landform. For example, some mountains are low. Their sides are gentle and people can climb them easily. Other mountains are high and steep. Only mountain climbers can climb them. It is very cold at the tops of the tallest mountains.

Look at the pictures of a valley and a canyon. These landforms are usually formed by flowing water. Weathering and erosion by rivers slowly wore away the soil and rock. This happens day after day. Slowly, over time, Earth's landforms are changing. But most of these changes are too slow for people to notice.

Digital Library

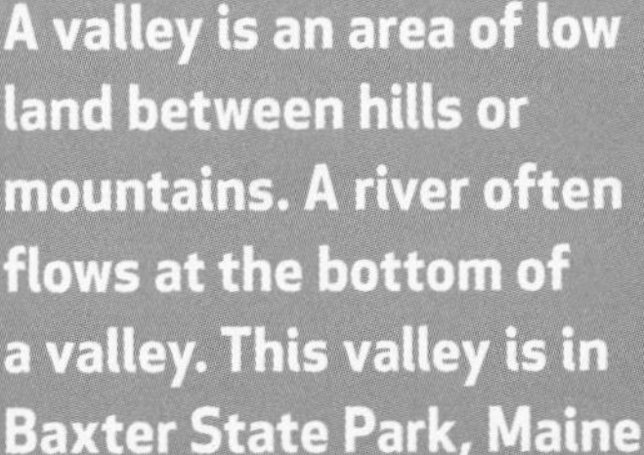

VALLEY

A valley is an area of low land between hills or mountains. A river often flows at the bottom of a valley. This valley is in Baxter State Park, Maine.

CANYON

A canyon is a deep valley between cliffs. Most canyons have a river at the bottom. This canyon is part of the Grand Canyon in Arizona.

Before You Move On

1. Name landforms you have seen.
2. How are valleys and canyons alike? How are they different?
3. **Draw Conclusions** Does any landform always stay the same? Why or why not?

NATIONAL GEOGRAPHIC

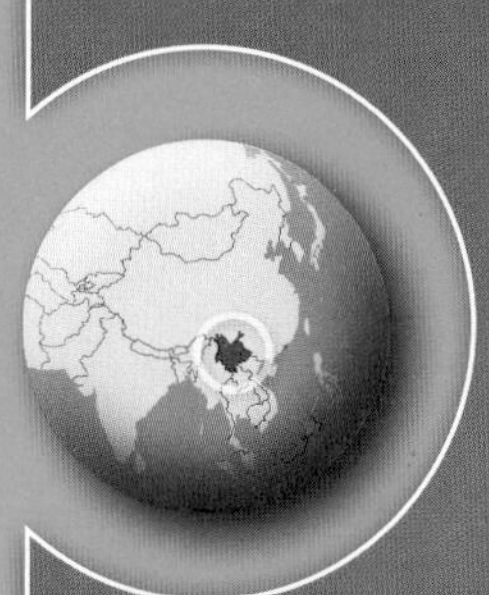

TERRACE FARMING
IN YUNNAN PROVINCE, CHINA

Yunnan (YÜ-nahn) Province is in southern China, in Asia. China is almost as big as the United States. China has many types of landforms, including mountains and hills. Soil often erodes from the sides of these landforms. Soil erosion is a big problem in the Yunnan Province. Farmers have a hard time growing food when their land is eroding away.

The side of this hill has been changed into stair steps of small fields that wrap around the hill.

To make farming possible, farmers in Yunnan Province use terraces. They look like stair steps. Terraces are raised flat areas of soil or other materials built across the slopes of hills. Terraces stop soil from eroding. They keep water near the plants that need it. Terraces keep the water from running downhill and causing more erosion.

Farmers tend their crops on these terraces.

Conclusion

Earth's surface is always slowly changing. Weathering breaks down or wears away rocks and other materials. Erosion carries small pieces of these weathered materials from one place to another. The eroded materials are deposited in a new place. These slow changes shape Earth's many landforms.

Big Idea Weathering, erosion, and deposition are slowly changing Earth's surface.

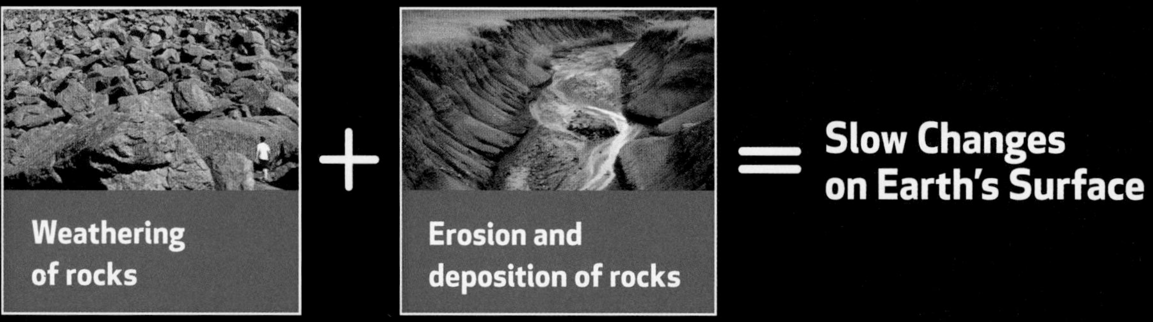

Vocabulary Review

Match the following terms with the correct definition.

A. weathering
B. erosion
C. glacier
D. deposition
E. landform

1. The laying down of rocks and soil in a new place
2. A natural feature on Earth's surface
3. The breaking apart, wearing away, or dissolving of rocks
4. A huge area of ice that moves slowly over Earth's surface
5. The picking up and moving of rocks and soil to a new place

Big Idea Review

1. **List** What are some major landforms?

2. **Recall** How does ice weather rock?

3. **Explain** How does a glacier change the land?

4. **Cause and Effect** Will a fast moving river or a slow moving river weather rock more quickly? Explain your answer.

5. **Generalize** Why are all landforms affected by weathering, erosion, and deposition?

6. **Predict** What might happen to a house built right along the shore of an ocean, lake, or river?

Write About Earth's Surface

Draw Conclusions What is happening in this photo? How do you think the sea stacks formed? Write about what will happen to them over time.

Why does the land look the way it does? Ask a geomorphologist.

Margaret Hiza Redsteer is a geomorphologist. She studies the shape of the land and how it changes. That's what *geomorphologist* means. *Geo-* means "Earth," *morph-* means "shape," and *-ologist* means "person who studies."

Redsteer studies the layers of Navajo sandstone near Tuba City, Arizona.

Redsteer studies how land changes. She looks at how the weather affects the land over time. She also looks at how people can change the land due to farming and other activities.

Redsteer works mainly on lands in Arizona, New Mexico, and Utah. This is the location of the Native American people known as the Navajo. "I want to improve the lives of Native people," she says. "Native American communities need Earth science information in order to plan and grow."

TECHTREK
myNGconnect.com
Student eEdition
Digital Library

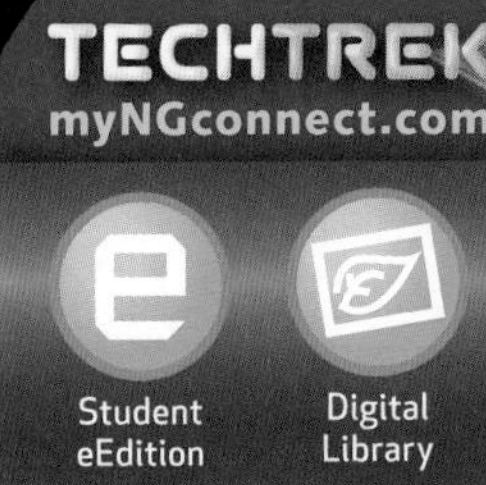

In these states, Redsteer takes samples of rocks, soils, sand, and dust. She also gets information from weather stations. She puts her information together to tell how far sand and soil have moved from one place to another.

Are you interested in being a geomorphologist? Like Redsteer, you would probably spend a lot of time outside "in the field." You might also spend time using equipment in a lab, writing research articles, and working with a team of other scientists. So besides studying science in school, it's important to work on writing skills. "Scientists always study, and they always keep learning," Redsteer says. "It's what makes it fun to be a scientist."

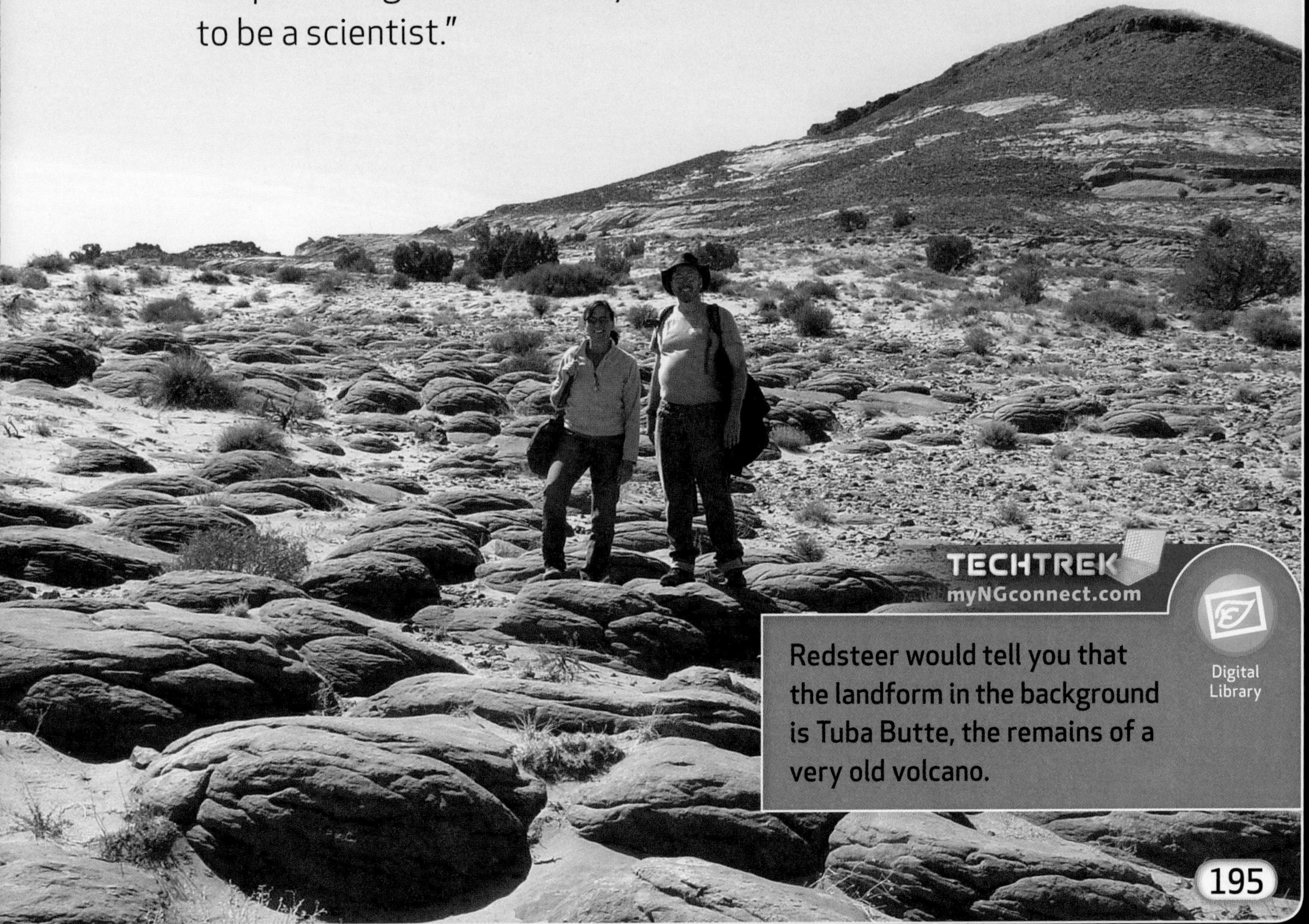

TECHTREK
myNGconnect.com
Digital Library

Redsteer would tell you that the landform in the background is Tuba Butte, the remains of a very old volcano.

NATIONAL GEOGRAPHIC

BECOME AN EXPERT

Weird and Wonderful Caves

Cave. This one word is enough to make some people want to put on a headlamp and go exploring. You might have visited a cave or seen people exploring one on a television program. How do these special places come to be? They are formed by **weathering** and **erosion**.

Sandstone caves form at the base of cliffs. Water loosens the particles that make up the cliff. Wind carries them away. The top of the cliff hangs over the cave.

weathering

Weathering is the breaking apart, wearing away, or dissolving of rocks.

erosion

Erosion is the picking up and moving of rocks and soil to a new place.

TECHTREK
myNGconnect.com
Student eEdition | Enrichment Activities | Digital Library

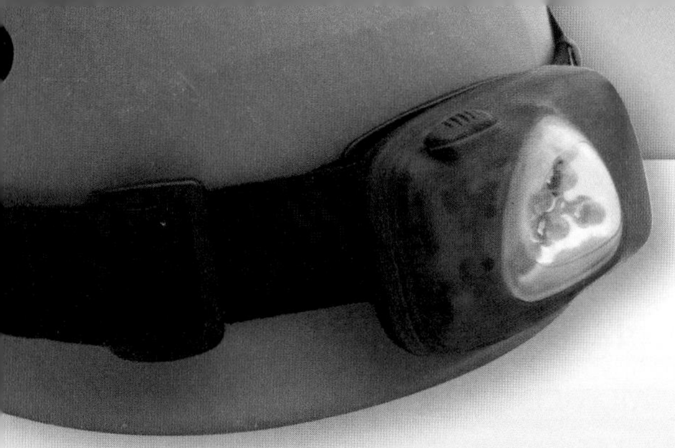

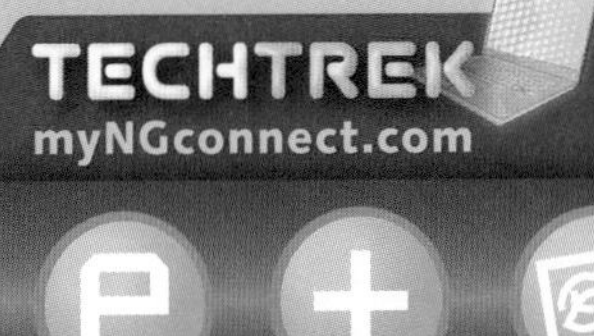

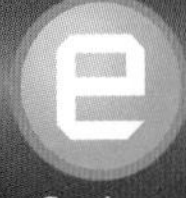

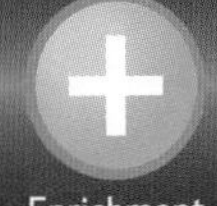

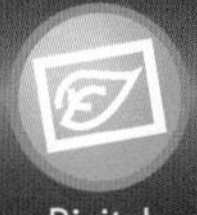

Sea caves form as waves strike cliffs on the coastline. What started out as a crack slowly becomes a cave. Rivers form caves in much the same way. When rain or a stream flows over the side of a sandstone cliff, it slowly wears away the soft rock and forms shallow caves.

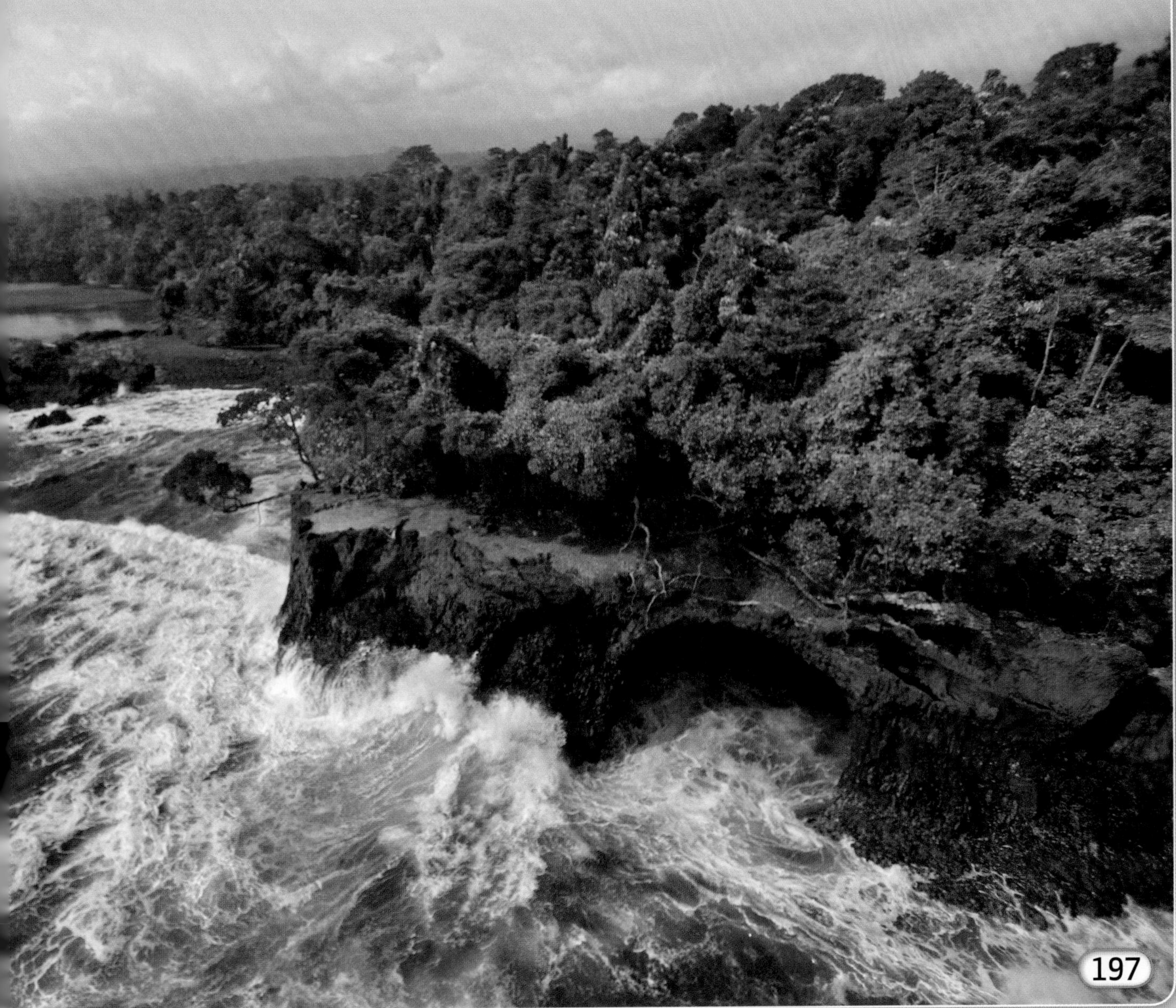

Caves can form almost anywhere, even in **glaciers**! A glacier cave forms when the sun melts ice on the surface of a glacier. The water runs into cracks in the glacier and flows under the glacier. As the water flows, it melts more ice, and a cave forms.

The Paradise Ice Caves in Mount Rainier's Paradise Glacier in Washington State were very popular with hikers. The caves fell as the glacier slowly moved.

glacier

A **glacier** is a huge area of ice that moves slowly over the land.

Many caves are limestone caves. These **landforms** are made by a kind of weathering called chemical weathering. Rain mixes with carbon dioxide gas and weathers limestone by dissolving it. The water erodes the dissolved limestone, and a cave slowly forms. Eventually, the water drains out of the cave.

TECHTREK myNGconnect.com

Enrichment Activities

HOW CAVES FORM IN LIMESTONE

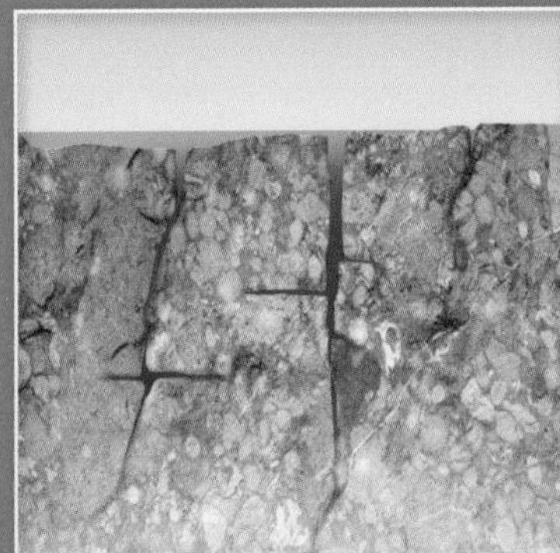

STAGE 1
Rainwater mixes with carbon dioxide gas. This mixture runs into cracks. The water dissolves the limestone.

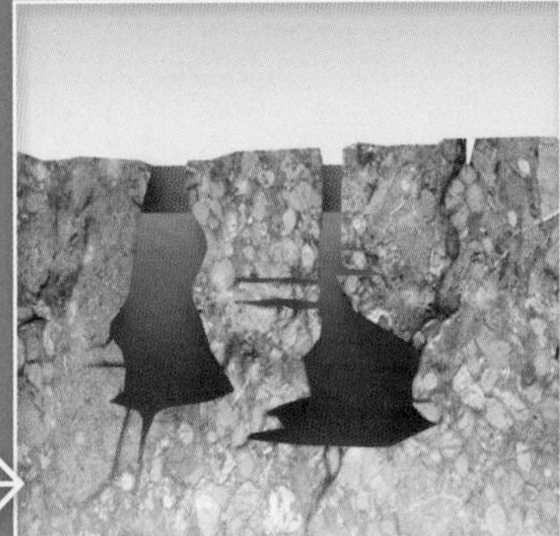

STAGE 2
Water keeps seeping into cracks and dissolving rock. Over thousands of years, cracks and holes get bigger.

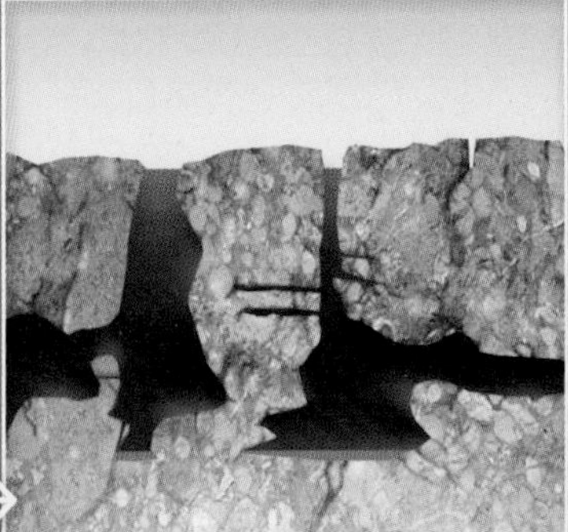

STAGE 3
Forces deep within Earth slowly push the land up. Water drains from the cave and it stops getting bigger.

landform

A **landform** is a natural feature on Earth's surface.

BECOME AN EXPERT

Many different features form in limestone caves. This happens because of **deposition**. Even though most of the water has left the cave, some water still seeps in. As the water drips into the cave, it deposits minerals on the roof of the cave.

Deposits are still forming in the Luray Caverns in the Shenandoah Valley of Virginia.

Stalagmites form under stalactites. They meet in the middle and form a column.

deposition

Deposition is the laying down of rocks and soil in a new place.

The minerals build up on the roof until they form a long cone. This is called a stalactite. Water also drips on the cave's floor and leaves minerals there. They build up to form a stalagmite. Many other amazing features form from minerals that come out of water that drips into a cave.

TECHTREK
myNGconnect.com

Digital Library

When stalactites first start to form, they are very thin. They are sometimes called "soda straws."

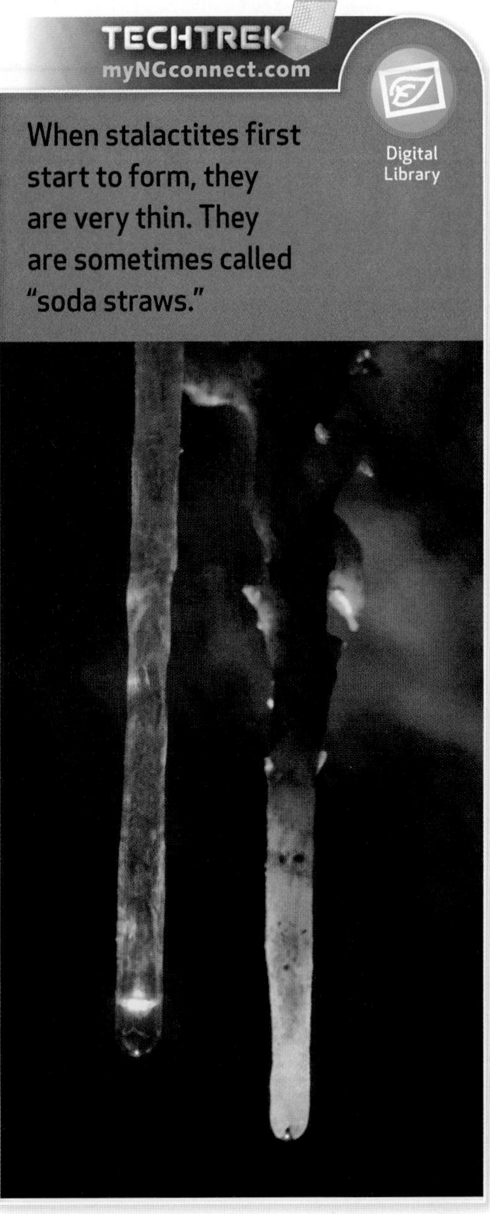

Mammoth Cave in Kentucky is the longest series of caves in the world. People have explored more than 580 km (360 miles) of Mammoth Cave. But scientists think a lot more of the cave remains to be discovered. Mammoth Cave began forming about 10 million years ago. It has stalactites, stalagmites, and many other features.

This large section of Mammoth Cave is called Booth's Amphitheater.

Some of these features look like flowing liquid, curtains, flowers, or even popcorn. They all start the same way. Water that's full of minerals flows into a cave, depositing minerals on the ceiling, floor, and walls. These formations take millions of years to build, and some break easily. So if you ever visit a cave like Mammoth, be careful.

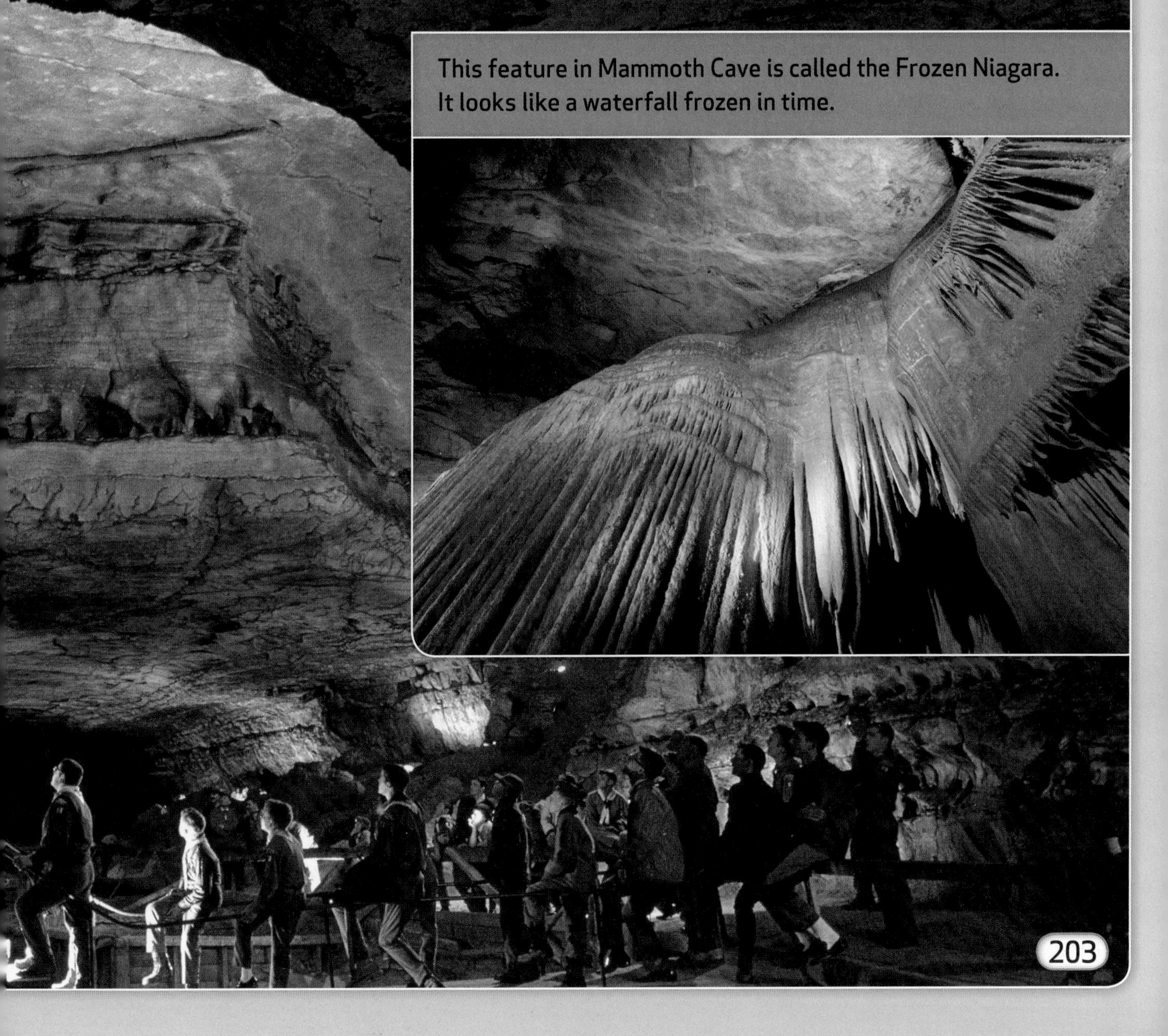

This feature in Mammoth Cave is called the Frozen Niagara. It looks like a waterfall frozen in time.

SHARE AND COMPARE

Turn and Talk How do weathering, erosion, and deposition each help form a cave? Form a complete answer to this question together with a partner.

Read Select two pages in this section. Practice reading the pages. Then read them aloud to a partner. Talk about why the pages are interesting.

my SCIENCE notebook **Write** Write a conclusion that tells the important ideas about what you learned about caves. State what you think is the Big Idea of this section. Share what you wrote with a classmate. Compare your conclusions.

my SCIENCE notebook **Draw** Draw a cave that you think might be interesting to explore. Combine your drawing with those of the classmates in your group. Connect the caves with passages so that it looks like one large series of caves. Explain with your group how the caves formed.

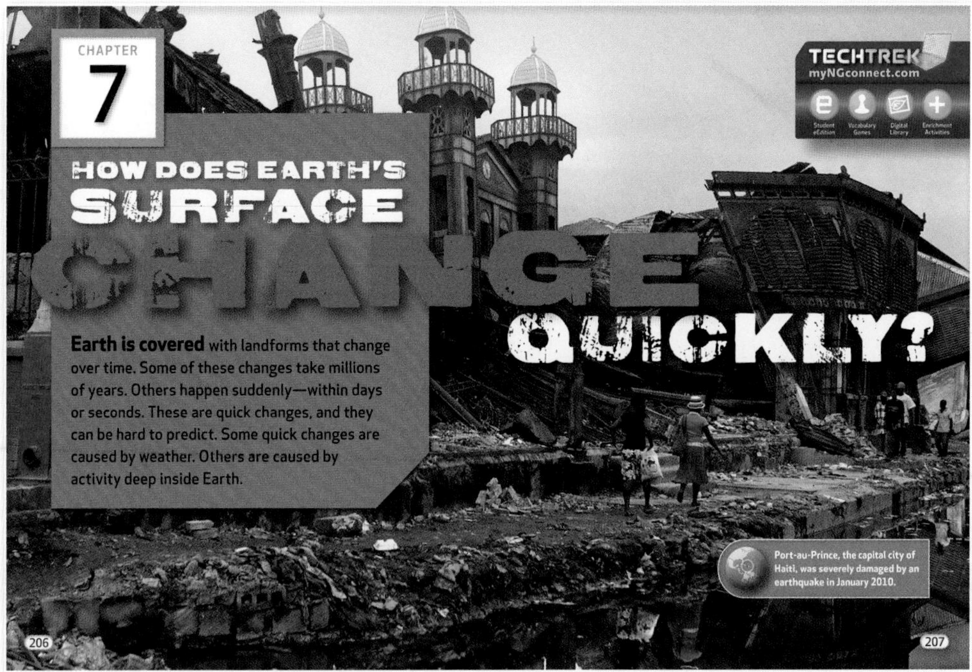

After reading Chapter 7, you will be able to:

- Recognize that earthquakes can change Earth's surface quickly. **EARTHQUAKES**
- Recognize that volcanoes can change Earth's surface quickly. **VOLCANOES**
- Recognize that landslides can change Earth's surface quickly. **LANDSLIDES**
- Recognize that extreme weather and other natural events can change Earth's surface quickly. **EXTREME NATURE**
- Science in a Snap! Recognize that earthquakes can change Earth's surface quickly. **EARTHQUAKES**

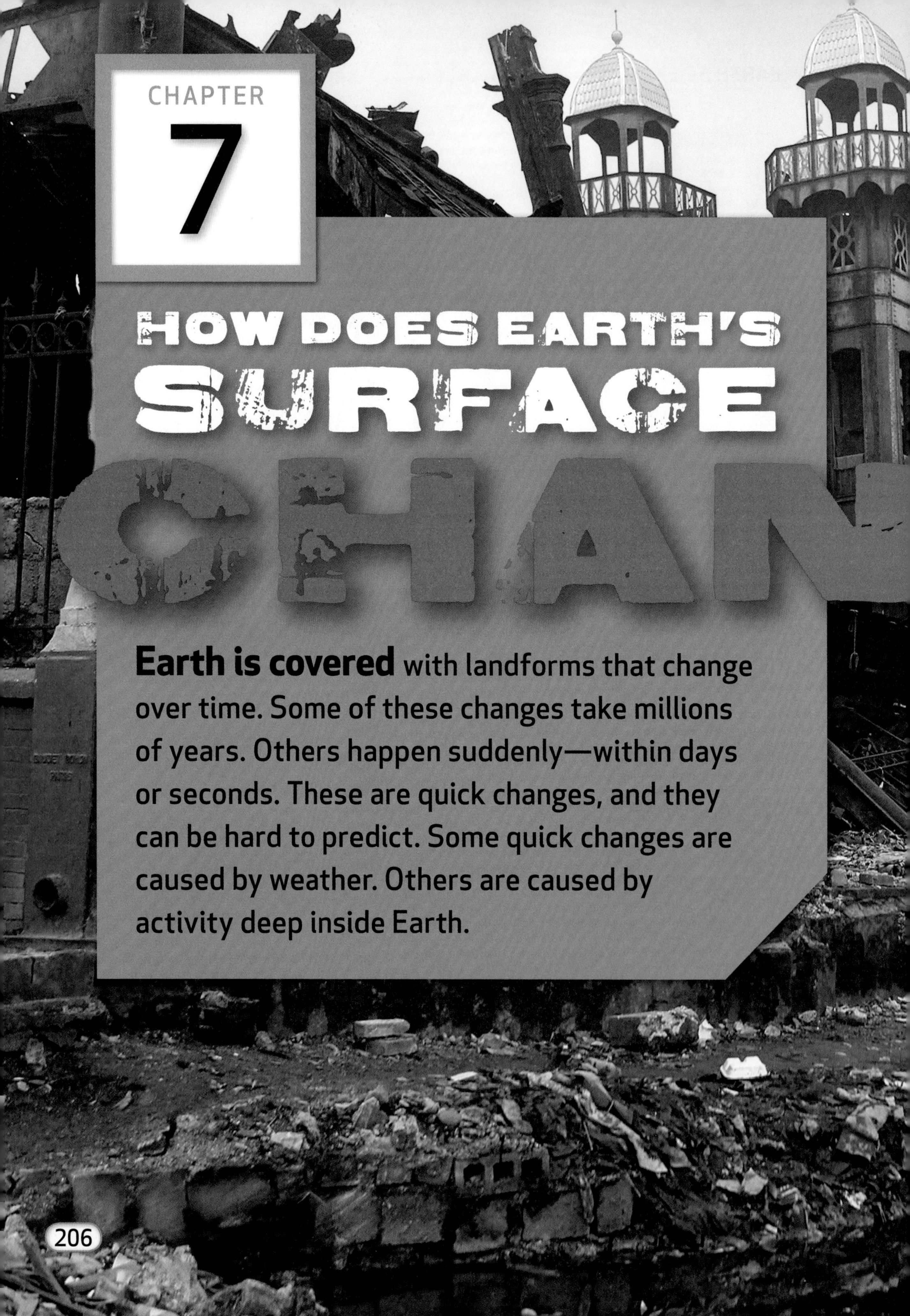

CHAPTER

7

HOW DOES EARTH'S SURFACE CHAN

Earth is covered with landforms that change over time. Some of these changes take millions of years. Others happen suddenly—within days or seconds. These are quick changes, and they can be hard to predict. Some quick changes are caused by weather. Others are caused by activity deep inside Earth.

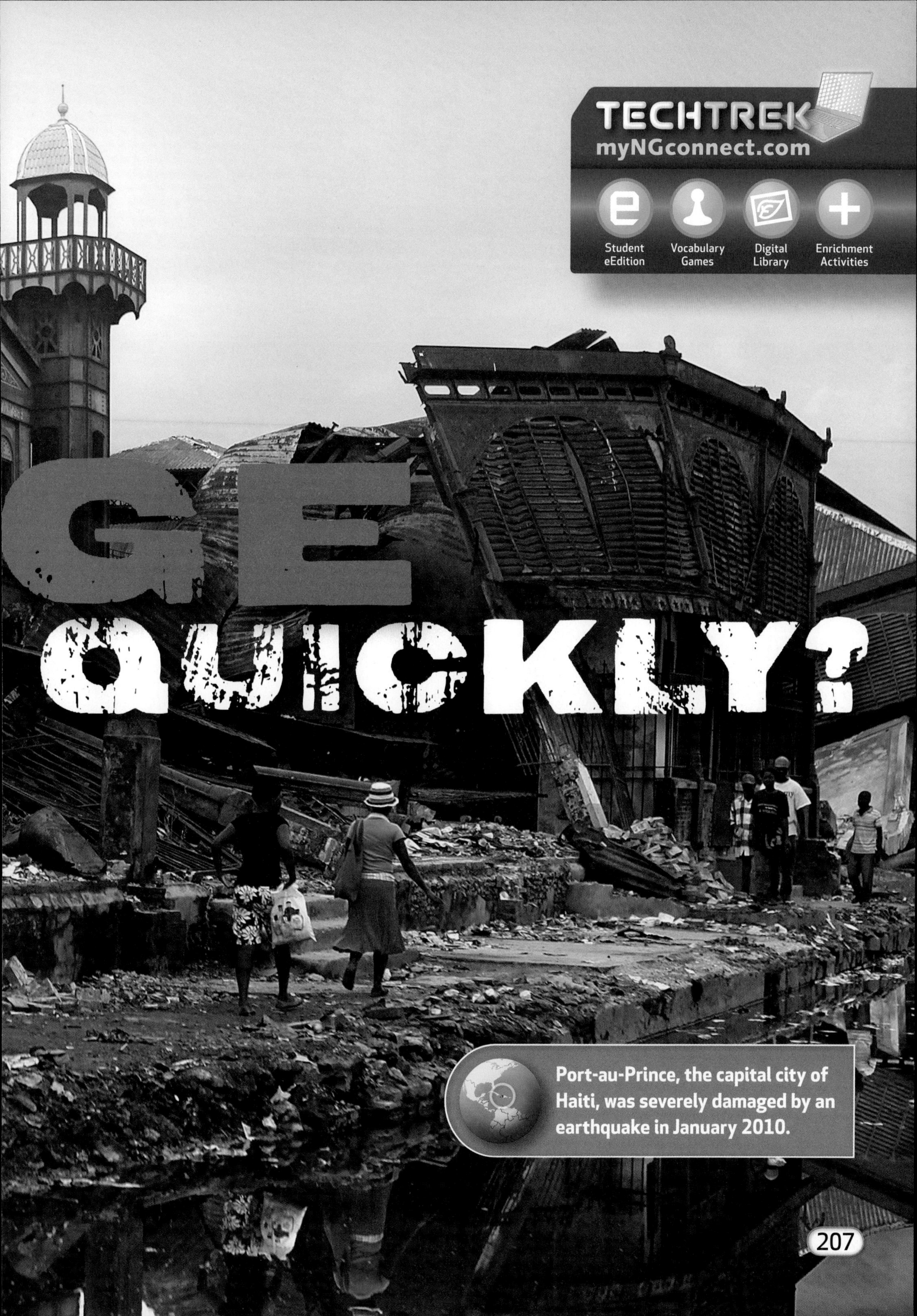

Port-au-Prince, the capital city of Haiti, was severely damaged by an earthquake in January 2010.

SCIENCE VOCABULARY

earthquake (URTH-kwāk)

An **earthquake** is the shaking of the ground caused by the movement of Earth's crust. (p. 210)

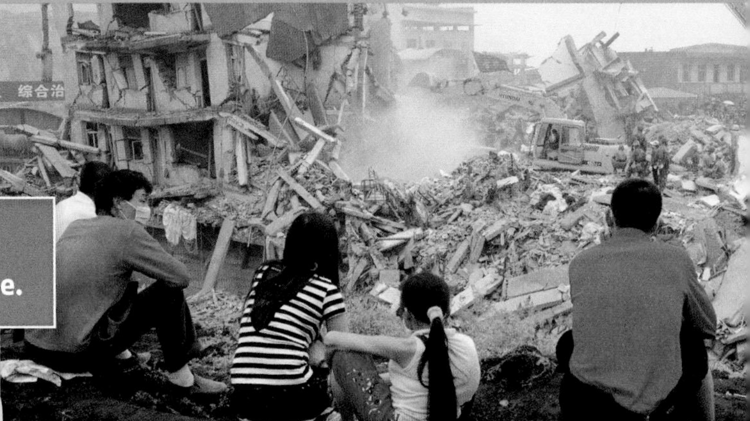

Earthquakes can cause cracks in Earth's surface.

plate (PLĀT)

A **plate** is a huge piece of Earth that floats on a layer of liquid rock. (p. 211)

Earth's crust is separated into plates that move.

earthquake
(URTH-kwāk)

lava
(LAH-vah)

magma
(MAG-mah)

plate
(PLĀT)

volcano
(vol-KĀ-nō)

volcano (vol-KĀ-nō)

A **volcano** is an opening in Earth's crust through which lava, gases, and ash erupt. (p. 214)

A volcano can grow into a mountain over many years.

magma (MAG-mah)

Magma is melted rock below Earth's surface. (p. 215)

Magma can move to Earth's surface through volcanoes such as this one.

lava (LAH-vah)

Lava is melted rock that comes to Earth's surface through a volcano. (p. 215)

Lava flows out of this volcano.

Earthquakes

Imagine what it would be like if Earth's surface began to move. Buildings sway, roads crack, and bridges fall. Then, just as quickly the shaking stops. This is what a powerful **earthquake** may be like. An earthquake is the shaking of the ground caused by movement of Earth's surface.

Science in a Snap! Plate Movement

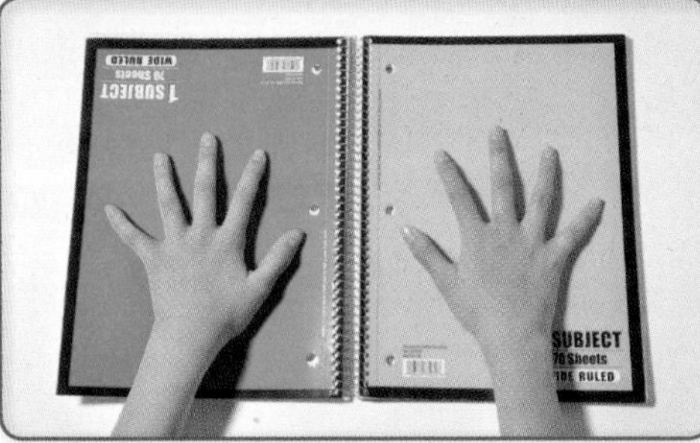

Place two spiral notebooks side-by-side so that the spirals are touching.

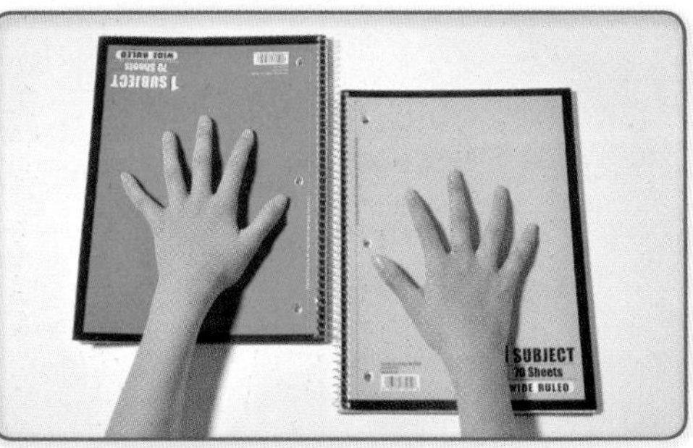

Place a hand firmly on top of each notebook and slowly push them in opposite directions.

How is this motion like that of plates? What happened as the notebooks scraped past each other?

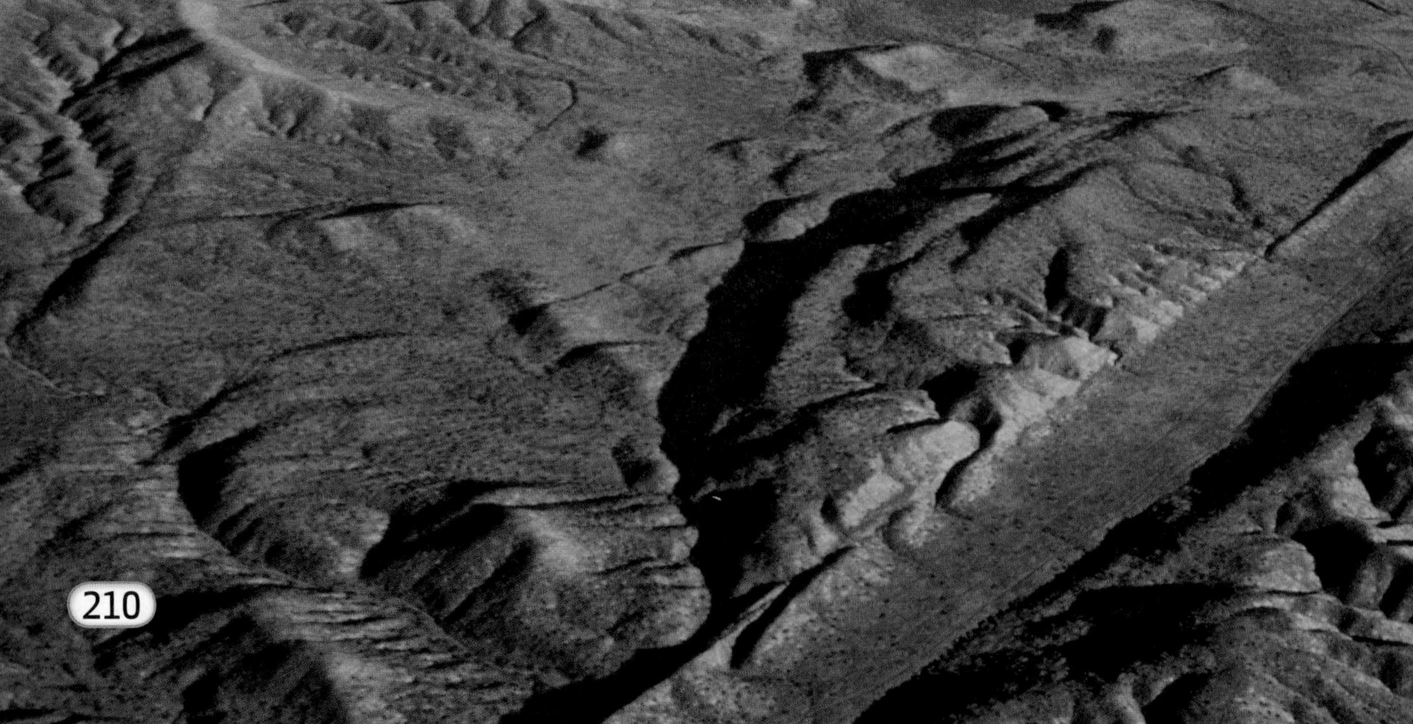

Earth's surface is divided up into several large sections called **plates**. Earth's plates float on a layer of melted rock. The melted rock causes the plates to move. The moving plates can push against each other. They can scrape past each other, too. Plates can even move away from each other. Some of these plate movements cause earthquakes.

This is the San Andreas Fault in California. It shows where two plates come together.

Earthquakes happen when plates lock together and energy builds up. When energy is released, the ground shakes. There are many earthquakes everyday around the world. Most of these are too small to be felt by people. They cause little or no damage.

Some earthquakes are very powerful. It can take a very long time for enough energy to build up to cause a powerful earthquake.

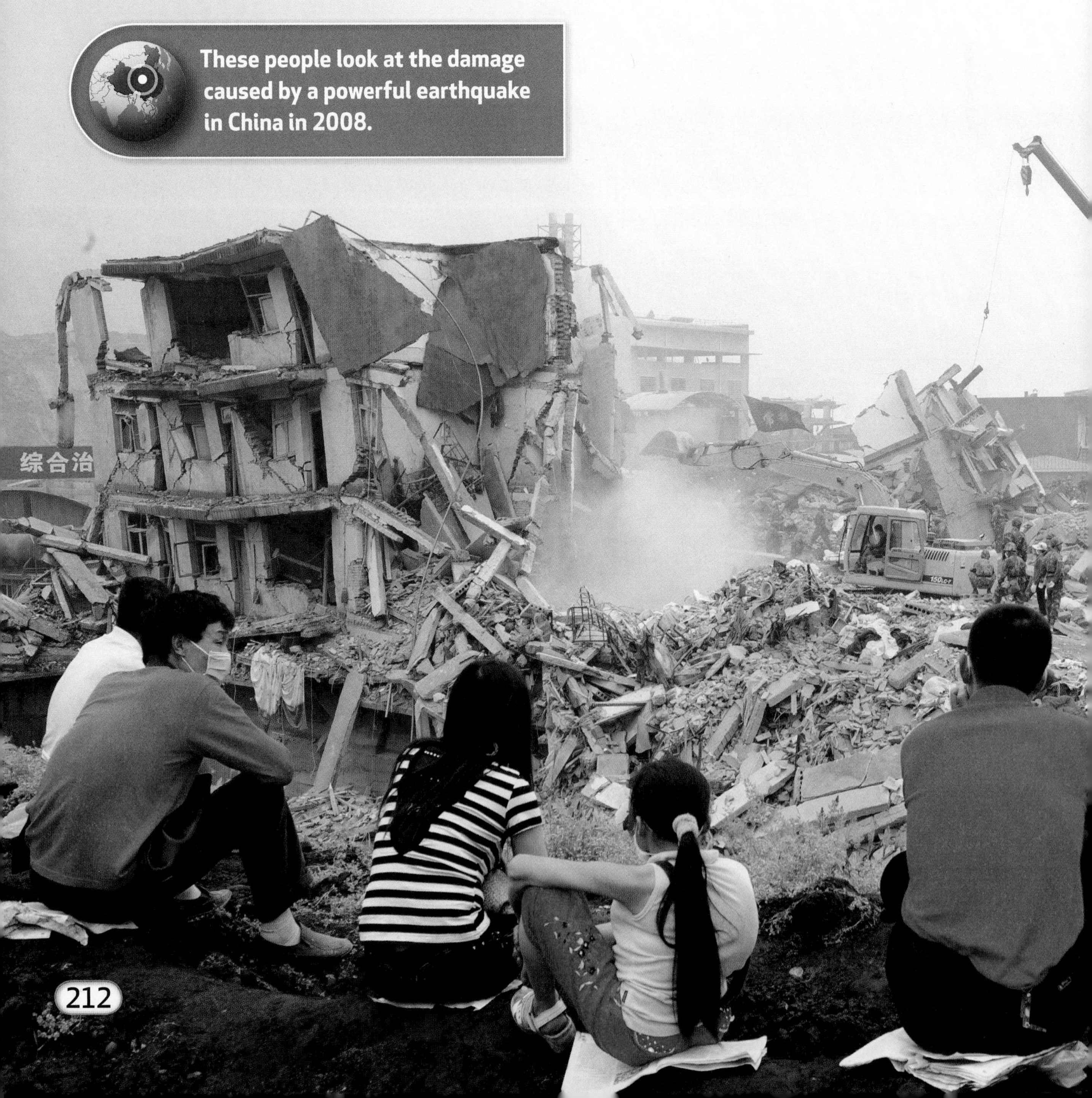

These people look at the damage caused by a powerful earthquake in China in 2008.

Most earthquakes do not cause huge changes to Earth's surface. They may crack the land or cause a landslide. However, in a few years the changes caused by the earthquake are no longer noticeable. But earthquakes can cause much damage to buildings and other structures where people live. It can take a long time to rebuild homes and businesses.

An earthquake caused the gash that cuts through these rice fields in Japan.

Before You Move On

1. List three ways that plates can move.
2. Explain what causes earthquakes.
3. **Infer** Why do you think earthquakes are more dangerous to people than to living things in the wild?

Volcanoes

Openings in Earth's surface can form near the edges of plates. A **volcano** can form at one of these openings. Melted rock, ash, and gases erupt or move up through the opening. Over time, this material builds up and creates new land.

Mt. Mayon is the most active volcano in the Philippines.

Magma is hot, melted rock that is under Earth's surface. Magma that reaches Earth's surface is called **lava**. When a volcano erupts, it covers the land with lava. The lava hardens into new land or rock as it cools. Over time the volcano gets bigger. Some volcanoes erupt often and can build very quickly. Others erupt less often and take many years to build.

TECHTREK
myNGconnect.com

Enrichment Activities

Magma rises up a volcano and erupts as lava. Ash and gases can erupt into the air.

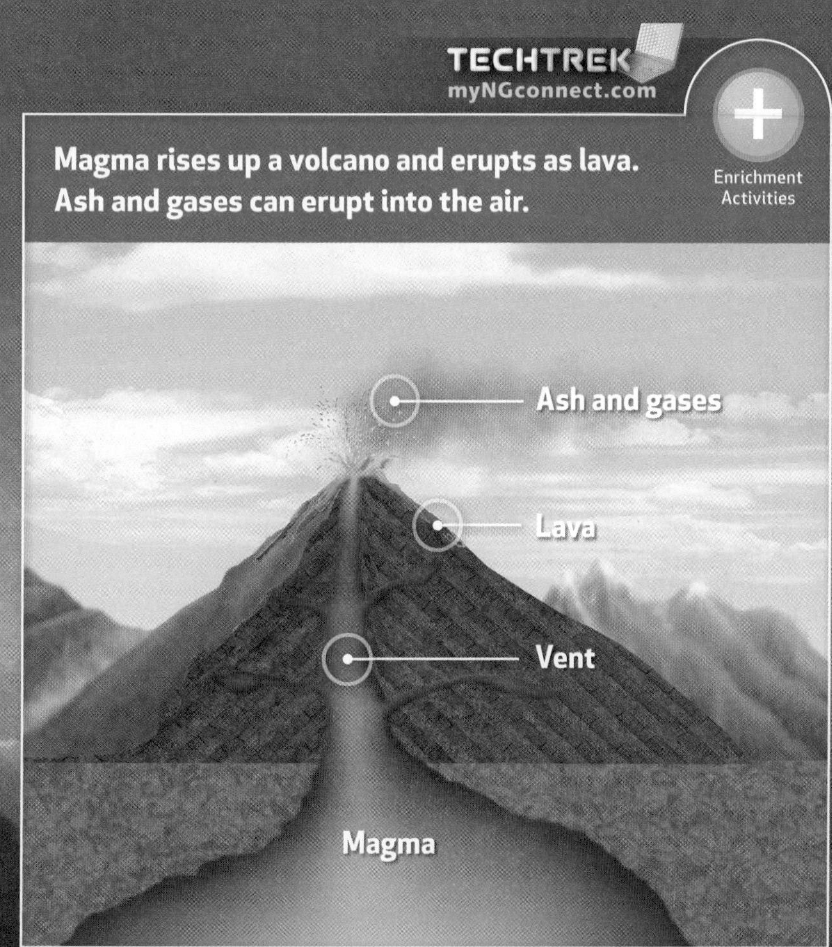

Volcanoes can change Earth's surface in years, days, or minutes. In a powerful eruption lava and ash cover the land. Harmful gases can fill the air. Ash darkens the sky. Ash can cover huge areas when volcanoes erupt, killing plants and animals. People have to move away. Sometimes whole towns are destroyed.

This valley on Montserrat was lush and green before the Soufriere Hills volcano erupted.

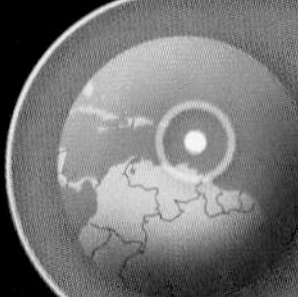

Montserrat is an island in the Caribbean. The Soufriere Hills volcano is on Montserrat. It has been erupting on and off since 1995.

Eruptions can be good, too. Gold and silver have been found in new volcanic rocks—so have opals and diamonds. Some underwater volcanoes form new islands. The minerals in volcanic ash improve the soil. Forests and crops grow better. It may be many years before a certain volcano erupts again. Living things have time to recover.

Here is the same valley after the eruption. The land is blanketed with ash.

Before You Move On

1. Define what a volcano is.
2. Contrast magma with lava.
3. **Evaluate** How can volcanic eruptions be both good and bad for the people living near volcanoes?

Landslides

Landslides happen when rock and soil slide down a slope. If enough rock falls, the shape of hills and mountains will change. In a landslide, gravity is the force that pulls rock down. But earthquakes, volcanoes, and heavy rains usually start them. How? By loosening large pieces of rock.

Landslides leave scars or bare spots on mountains.

Like earthquakes, landslides cause damage when they happen near people. Houses, schools, and other structures can be crushed under tons of rock and soil. Landslides can cover roads and bridges, making it harder for people to get help.

This landslide in El Salvador was started by an earthquake. It buried hundreds of homes.

Before You Move On

1. Identify some causes of landslides.
2. Explain how landslides change Earth's surface.
3. **Analyze** What part does gravity play in causing landslides?

Extreme Nature

Extreme weather can change Earth's surface in hours or days. Storms also can cause other events such as floods and fires.

Thunderstorms Have you ever been caught in a thunderstorm? Thunderstorms produce heavy rains, strong winds, and lightning. Some cause landslides and damage property. But thunderstorms also bring rain, which is good for plants, animals, and people.

Heavy Rain

Strong Winds

Lightning strikes during a thunderstorm over the city of Jerome, Arizona.

Blizzards If you live where winters are cold, chances are you have experienced a blizzard. Blizzards are winter storms with heavy snowfall and high winds. Like thunderstorms, blizzards can damage property and cause landslides. In spring, melting snow from blizzards can fill rivers and provide water for crops.

TECHTREK
myNGconnect.com

Heavy Snowfall

Digital Library

Hurricanes Hurricanes are tropical storms that form over warm ocean water. All hurricanes have high winds and heavy rains. They are nature's most powerful storms. When a hurricane reaches land, strong winds can damage homes and buildings. Large waves can cause floods and destroy beaches. But hurricanes move slowly. This usually gives people living near the ocean time to leave their homes.

This photo from space shows the swirling shape of a hurricane.

Hurricanes have wind speeds of at least 119 kilometers (74 miles) per hour. That's faster than a car speeding down a highway.

Tornadoes If you ever look at a stormy sky and see a funnel-shaped cloud, find shelter quickly. Tornadoes are dangerous clouds of swirling wind. Unlike hurricanes, tornadoes form suddenly and can move quickly. The strongest ones can damage almost anything in their path, including buildings, roads, and homes. Tornadoes can form over land or water, at any time of year.

This tornado in South Dakota had wind speeds greater than 333 kilometers (207 miles) per hour. That's strong enough to uproot trees and toss cars into the air.

Floods Have you ever seen a street or field covered with water? A flood is water that overflows onto dry land. Big floods can destroy cities and towns. Some floods are caused by hurricanes pushing ocean water onto land. Most floods result when a river overflows its banks. Floodwater can be helpful, too. The water can leave behind particles called sediment that make the soil better for plants.

This flood in Oklahoma happened when heavy rain caused a creek to rise almost 8 meters (25 feet).

Fires A wildfire is an uncontrolled fire in a wooded area. Lightning from storms can start wildfires. Some burn for weeks. Hot, windy, dry weather can spread fire quickly across huge areas of land. People's homes sometimes are destroyed. Many of the plants in a wildfire's path are burned up. However, wildfires can be good for forests. The burned material left over makes soil richer. Many plants have seeds that can survive the high heat.

Lightning started this forest fire in Custer State Park, South Dakota.

Before You Move On

1. Heavy thunderstorms cause a river to overflow. Streets fill with water. Which kind of extreme event is this?
2. Name one other kind of extreme weather or natural event. Suggest how it could change Earth's surface.
3. **Infer** How might a thunderstorm cause a landslide?

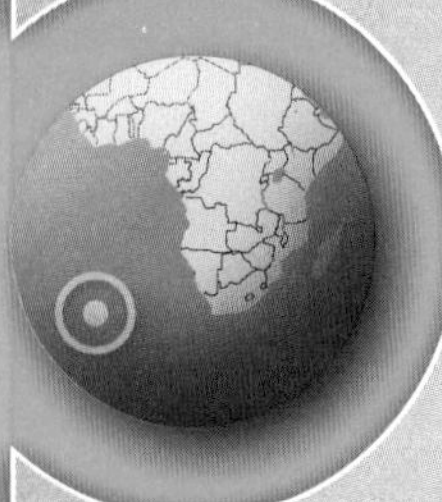

LIVING WITH A VOLCANO
TRISTAN DA CUNHA

How do you live with a volcano at your doorstep? The people of Tristan da Cunha are used to their volcano. Tristan da Cunha is a small volcanic island in the Atlantic Ocean. Families farm the land and raise animals. They make money by selling lobsters, stamps, and postcards to people all over the world.

This is the island from above. You can see that the island is the top of a very big volcano!

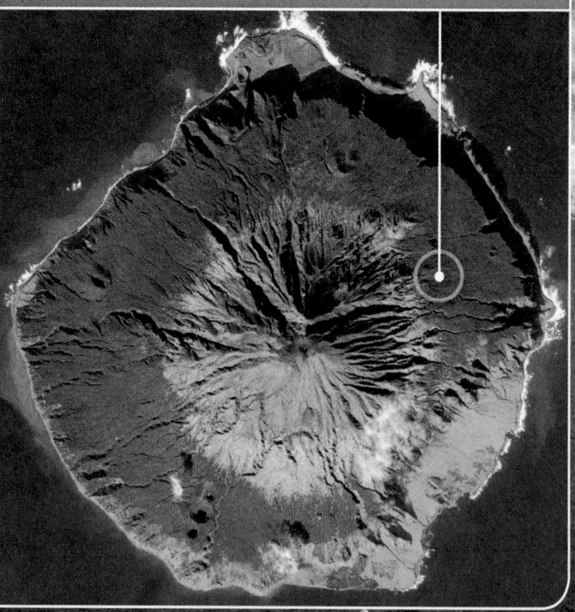

The village looks small on the volcano island.

In 1961, magma inside the volcano pushed toward the surface. This caused earthquakes and landslides. The volcano was going to erupt! All of the villagers had to leave. But the Tristan people missed their quiet life on the island. Two years later, almost all of them returned home.

Tristans return to their homes after the 1961 eruption.

Men repair a road that had been covered by lava from the volcano.

Once home, Tristans saw that the damage was not as bad as expected. The fishing harbor was destroyed by lava. But most of their farm animals were still alive. The Tristan people repaired their houses and roads. They rebuilt the harbor, which was necessary for catching lobster. Their stamps and postcards became more popular than ever. Within a few years, life on the island was back to normal.

People work together to repair a roof on a neighbor's house.

Tristan da Cunha island is still an active volcano. That means it could erupt again at any time. In July 2004, a cluster of tremors, or small earthquakes, occurred under the volcano. Earthquakes are one sign that an eruption could happen soon. So far, this hasn't happened. Tristans don't seem too worried either. Apart from the tourists, life on their island remains peaceful and quiet.

Two hikers enjoy the view from the side of the volcano on Tristan da Cunha.

Conclusion

Earthquakes, volcanoes, landslides, and weather events can change Earth's surface. Earthquakes can cause the land to crack or rise. Volcanoes create new land. Landslides and extreme weather can also cause changes to the land. These changes may be harmful or helpful to living things.

Big Idea Earthquakes, volcanoes, landslides, and extreme weather can change Earth's surface quickly.

QUICK CHANGES TO THE EARTH'S SURFACE CAN BE CAUSED BY:

Volcanic Eruptions

Earthquakes

Landslides

Extreme Weather

Vocabulary Review

Match the following terms with the correct definition.

A. earthquake
B. lava
C. plate
D. magma
E. volcano

1. A large section of Earth's crust that slowly moves
2. An opening in Earth's crust through which lava, gases, and ash erupt
3. Shaking of the ground caused by the movement of Earth's crust
4. Melted rock that comes to Earth's surface
5. Melted rock below Earth's surface

Big Idea Review

1. **Recall** Which can make new land—a landslide or a volcanic eruption?

2. **Describe** Describe what happens during a powerful volcanic eruption.

3. **Explain** How can plate movement cause earthquakes?

4. **Cause and Effect** What causes a flood, and how does it affect the land?

5. **Predict** Suppose you live on the slope of a mountain. Small earthquakes are common there. What are you in danger of, other than earthquakes?

6. **Generalize** What changes are helpful to living things? Explain how they are helpful.

Write About Earth's Surface

Infer What has happened in this photo? Write about how it changed Earth's surface.

CHAPTER 7

EARTH SCIENCE EXPERT: GEOLOGIST

What's it like to study changes in Earth's surface? Ask geologist Anne Egger!

Geologist Anne Egger

Do you enjoy being in nature? Are you interested in learning how Earth changes? Then you might want to become a geologist like Anne Egger!

Q: As a geologist, what do you do?

I teach college students about Earth science. I also do research about Earth. This involves figuring out Earth's history in areas where continents are being pulled apart.

Egger takes her class on a field trip to Montara Beach, California. The students learn about a landslide that happened nearby.

Q: What is your day like?

There's no such thing as a typical day, which is why I love my job. In the summer, I usually spend a few weeks doing research. This is a wonderful time. I get to combine my love of hiking with learning about science.

Q: How can kids prepare to do what you do?

A lot of people might say to take as much math and science as you can. My advice is to follow your interests. They might lead somewhere unexpected.

Q: What are you most proud of?

Showing people that studying Earth science is fun and interesting. My greatest success has been giving people a new outlook on their home planet.

Egger wears instruments on her back. They measure Earth's magnetic forces and map her location.

Geologists use all kinds of tools to study Earth. This big hammer can make vibrations in the ground. Egger measures how the vibrations travel through the rock.

NATIONAL GEOGRAPHIC

BECOME AN EXPERT

Cities in Ash

One of the world's most dangerous **volcanoes** rises above the Bay of Naples in Italy. Its name is Mount Vesuvius. Mount Vesuvius last erupted in 1944, but it was a small eruption. So why is Mount Vesuvius feared?

About 1900 years ago, Mount Vesuvius had one of the biggest eruptions ever recorded. Two Roman cities—Pompeii and Herculaneum—were buried under ash and rock. Could Mount Vesuvius erupt like that again?

Today, many more people live near Mount Vesuvius. Another big eruption would be deadly.

volcano

A **volcano** is an opening in Earth's crust through which lava, gases, and ash erupt.

TECHTREK
myNGconnect.com
Student eEdition
Digital Library

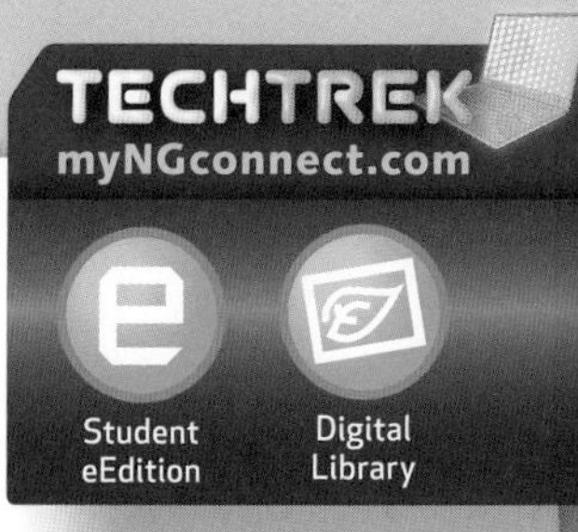

Mount Vesuvius is one of several volcanoes in Italy. These volcanoes are found where Earth's **plates** push against one other. Scientists watch Mount Vesuvius closely. They also study the history of Pompeii and Herculaneum. They want to learn all they can about what happened. The more scientists know, the more lives they can save the next time Mount Vesuvius erupts.

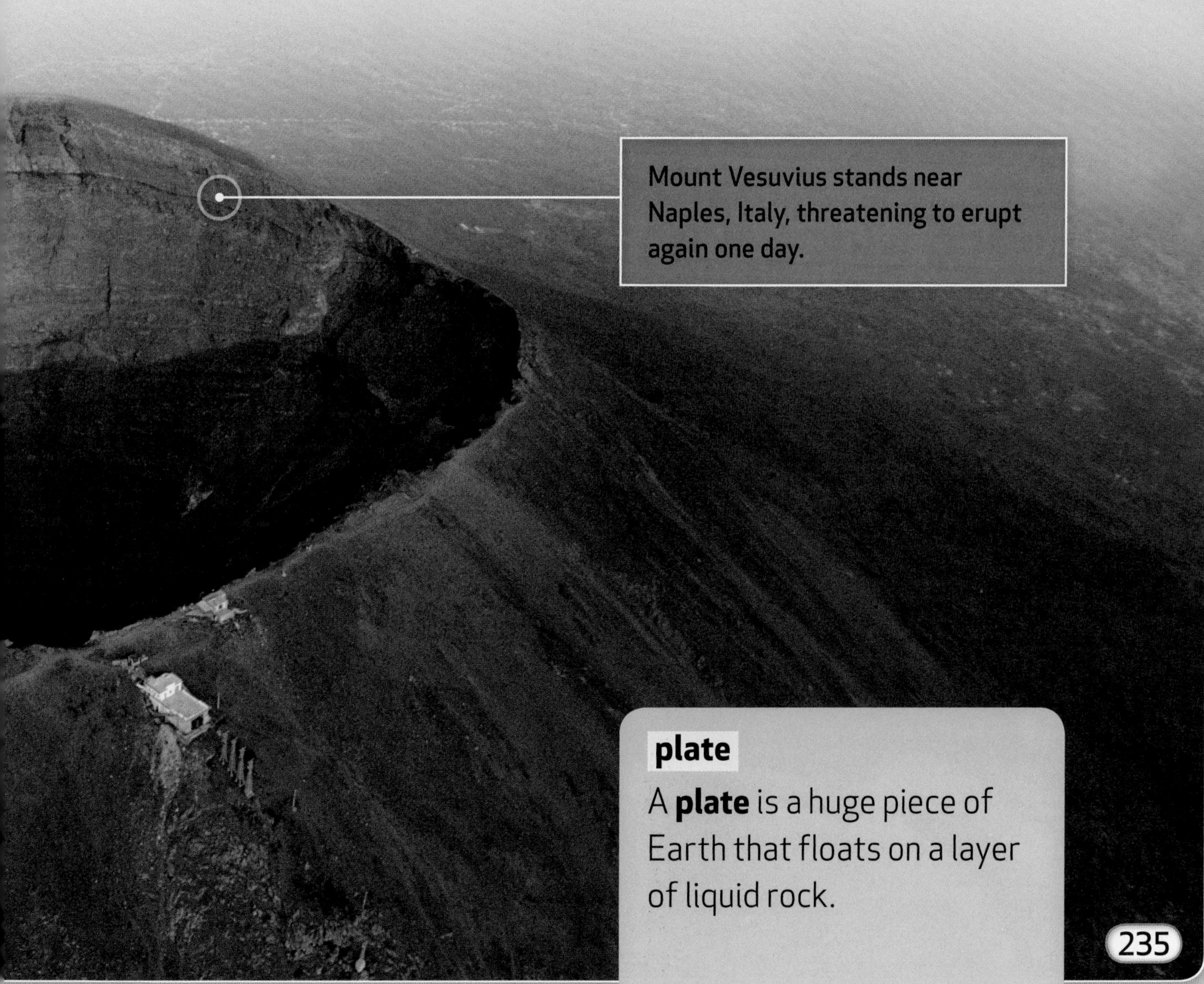

Mount Vesuvius stands near Naples, Italy, threatening to erupt again one day.

plate

A **plate** is a huge piece of Earth that floats on a layer of liquid rock.

Pompeii

Pompeii was a city much like any other. There were theaters, schools, and businesses. Then Mount Vesuvius erupted. In about one day, the city disappeared. Pompeii was buried under almost three meters (10 feet) of rock and ash. **Lava** flows added another three meters on top of that! Once buried, Pompeii was preserved and hidden from the world. More than 1700 years passed before it was discovered.

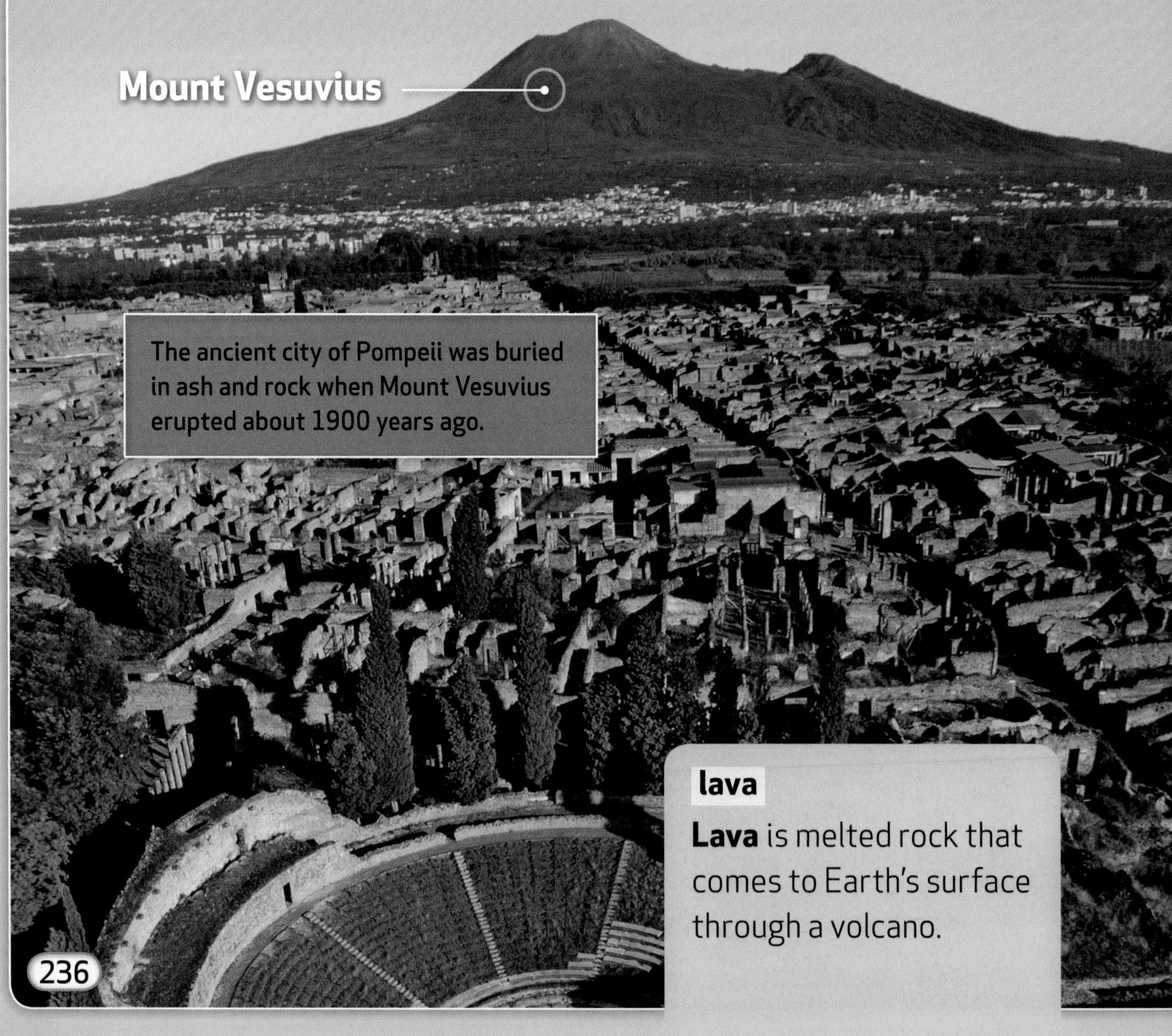

The ancient city of Pompeii was buried in ash and rock when Mount Vesuvius erupted about 1900 years ago.

lava

Lava is melted rock that comes to Earth's surface through a volcano.

Scientists and historians think that about 20,000 people lived in Pompeii. It was a wealthy city. Sometimes there were small earthquakes. But the people had never seen Mount Vesuvius erupt. Nobody thought that could happen. Yet, Mount Vesuvius must have erupted long before people lived there. That is why the land around Pompeii was good for growing crops. Ash from volcanoes can improve soil.

People in Pompeii had pets. This sign told people to beware of the dog.

TECHTREK myNGconnect.com

Digital Library

Wagons and other traffic once filled this street in Pompeii. Can you find the grooves, or ruts? They were made to keep the wagons level and on track.

Herculaneum

Not far from busy Pompeii was the quiet town of Herculaneum. About 5,000 Romans lived in Herculaneum. Its clear view of the ocean made this town a popular vacation spot. But Herculaneum was even closer to Mount Vesuvius than Pompeii.

An artist imagines what Herculaneum looked like before the eruption.

The eruption of Mount Vesuvius buried Herculaneum under more than 15 meters (50 feet) of mud and ash. That's about the height of a five-story building! The lava flows were much hotter than those of Pompeii. Hundreds of people did not escape in time.

The ancient city of Herculaneum was buried in mud and ash in the eruption.

Uncovering the Lost Cities

Scientists began digging up Pompeii and Herculaneum about 250 years ago. They wanted to learn more about human history. But digging carefully takes a very long time. Scientists today are still digging.

In the 1860s, one of these scientists found strange spaces in the rock. He realized something important—the spaces had been left there by human bodies! So he filled the spaces with plaster. When the plaster hardened, it became a model of the person's body.

These coins were discovered in Pompeii.

A plaster model of someone who lived in Pompeii and died in the eruption. The plaster preserves the shape of the person's body.

Herculaneum is better preserved than Pompeii. Why? Because the ash, gas, and rock that buried it was so hot. The ancient city is preserved in time.

In the 1980s, scientists discovered human skeletons at Herculaneum. It is rare to find skeletons from ancient Rome. Scientists studied the bones. They were able to learn more about the way people lived 1900 years ago.

A bakery, and even a loaf of bread, were preserved.

A scientist works carefully to uncover a skeleton at Herculaneum.

There are still things to learn from Pompeii and Herculaneum. Today the cities are mostly uncovered. People can walk down the streets and see the buildings. It is like walking back in time almost two thousand years.

Scientists believe that Mount Vesuvius will erupt again. The number of **earthquakes** has increased in recent years. Small earthquakes are one sign of a possible eruption. They happen when **magma** inside the volcano moves.

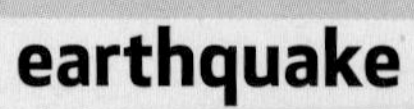

An **earthquake** is the shaking of the ground caused by the movement of Earth's crust.

Magma

Magma is melted rock below Earth's surface.

Mount Vesuvius is the most closely watched volcano in the world. Scientists would have to predict an eruption in time to move 600,000 people to safety. The remains of Pompeii and Herculaneum remind people of what could happen if they are not prepared.

Mount Vesuvius towers over the ruins of Pompeii.

SHARE AND COMPARE

Turn and Talk How are the ruins of Pompeii and Herculaneum the same, and how are they different? Form a complete answer to this question together with a partner.

Read Select two pages in this section. Practice reading the pages. Then read them aloud to a partner. Talk about why the pages are interesting.

my SCIENCE notebook **Write** Write a conclusion that tells the important idea about what you have learned about Pompeii and Herculaneum. State what you think is the Big Idea of this section. Share what you wrote with a classmate. Compare your conclusions.

my SCIENCE notebook **Draw** Choose a picture from Pompeii or Herculaneum. Use it to draw a picture of what might have been happening before the eruption. Add a caption and labels to your picture. Put your picture together with those of your classmates to create a journal of life in those cities.

After reading Chapter 8, you will be able to:

- Recognize that most of Earth's surface is covered by water. **THE WATER PLANET**
- Recognize that water can change form through evaporation, condensation, and freezing. **FORMS OF WATER**
- Recognize that water is recycled by natural processes on Earth. **WATER MOVES AROUND EARTH**
- Recognize that evaporation, condensation, precipitation, run-off and groundwater are all elements in the water cycle. **WATER MOVES AROUND EARTH**
- Science in a Snap! Recognize that water can change form through evaporation, condensation, and freezing. **FORMS OF WATER**

CHAPTER

8

HOW DOES EARTH'S WATER MOVE AND CHANGE?

How have you used water today? You might not have rafted down a river. Still, chances are you washed with water and drank it. Maybe you swam in it. It might have poured down on you as rain. People use water every day.

Student eEdition

Vocabulary Games

Digital Library

Enrichment Activities

WATER CHANGE?

These rafters are on the Salmon River in Idaho. You can observe how water in the river moves and changes.

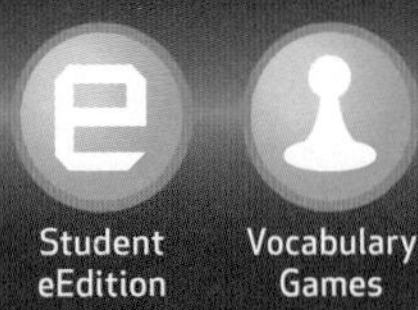

SCIENCE VOCABULARY

salt water (SAWLT waw-tr)

Salt water is water that has dissolved salts in it. (p. 251)

Most of Earth's water is the salt water in oceans.

fresh water (FRESH waw-tr)

Fresh water is water that has little or no salt. (p. 251)

Rivers and most lakes are fresh water.

freezing (FRĒZ-ing)

Freezing is the change from liquid to solid. (p. 252)

Ice forms when water's temperature goes below freezing.

condensation (kon-den-SĀ-shun)

evaporation (i-vap-uh-RĀ-shun)

freezing (frēz-ing)

fresh water (FRESH waw-tr)

salt water (SAWLT waw-tr)

water cycle (WAW-tr SĪ-kel)

evaporation
(i-vap-uh-RĀ-shun)

Evaporation is the change from a liquid to a gas. (p. 254)

The sun shines on Earth's surface, causing evaporation of surface waters.

condensation
(kon-den-SĀ-shun)

Condensation is the change from a gas to a liquid. (p. 254)

Condensation of water vapor in the air forms clouds.

water cycle (WAW-tr SĪ-kel)

The **water cycle** is the movement of water from Earth's surface to the air and back again. (p. 256)

The water cycle recycles Earth's water.

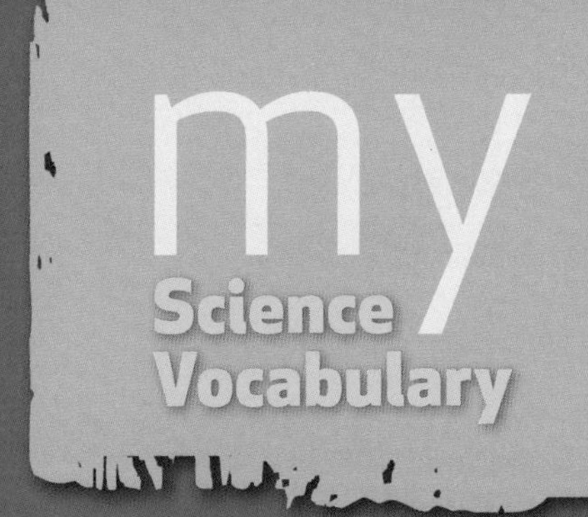

The Water Planet

This is Earth from space. What can you observe? It looks green, brown, white, and blue. The blue areas are liquid water. Most of Earth is blue. Why? Because water is almost everywhere on Earth's surface. Water is also underground and in the air. It is a big part of living things. Even your body is more than half water.

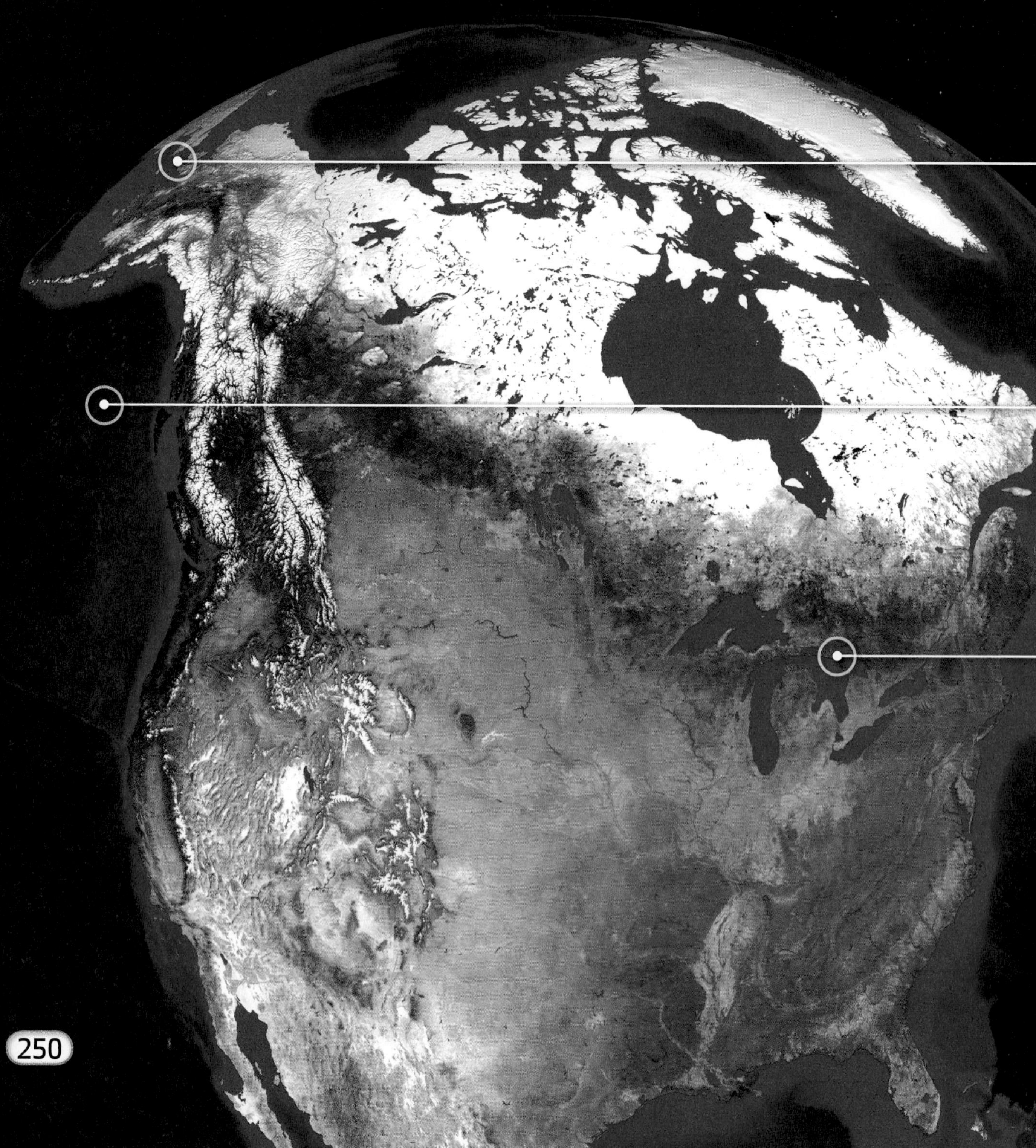

Most of Earth's water is salt water. Salt water has dissolved salt in it. It tastes salty. Earth's oceans are salt water. The water in most lakes and rivers is fresh water. Fresh water has little or no salt in it. Most of the water under the ground and in glaciers is also fresh water.

GLACIER
Glaciers and ice caps hold most of Earth's fresh water. But they still contain only a tiny fraction of Earth's water.

OCEAN
Oceans cover most of Earth's surface. Almost all of Earth's water is the salt water in the oceans.

LAKE
Lakes are mostly bodies of fresh water with land all around them. Lakes and rivers hold the smallest amount of Earth's water.

Before You Move On

1. What type of water covers most of Earth's surface?
2. How are salt water and fresh water different?
3. **Evaluate** Why is Earth called a water planet?

Forms of Water

Water can be a solid, a liquid, or a gas. Its form depends on its temperature. Most water on Earth is liquid. You see liquid water in rivers, ponds, and oceans. It falls as rain. You can also see water in its solid form, ice. Look at the icy pond. Why did the liquid water in the pond **freeze**, or turn from a liquid to a solid?

Some of the water in this pond got cold enough to freeze. Water freezes when its temperature is at or below 0°C (32°F).

Water can also exist as a gas. The gas is water vapor. Water vapor is all around you. You just can't see it. But you can feel it. Sometimes there is a lot of water vapor in the air on a hot day. The water vapor makes you feel sticky and sweaty.

This is the same bridge in summer. It is too warm for the pond to freeze.

Water changes form all the time. When you heat liquid water, it can **evaporate**. Evaporation is the change from a liquid to a gas. When you cool the gas, it can **condense**. Condensation is the change from a gas to a liquid. What can happen when you cool liquid water? It can freeze. Use the chart to compare what happens as water changes form.

Water can change form. It can be solid, liquid or gas.

Imagine you have a pot of boiling water. The water evaporates quickly because it is hot. It changes into a gas, which you can't see. The gas cools rapidly in the air. It condenses. The steam you see rising from the boiling water is condensation.

Science in a Snap! Changing Water

Pour warm water into a plastic soda bottle until it is about half full.

Place an ice cube over the opening in the top of the bottle.

What do you observe happening inside the bottle?

Condensation of water vapor near the ground forms clouds of fog.

Before You Move On

1. In which forms does water exist in nature?
2. What is the difference between evaporation, condensation, and freezing?
3. **Generalize** What must happen before water can change form?

Water Moves Around Earth

All living things use water to stay alive. People use it every day. Yet Earth's water never runs out. Why? Because Earth recycles its water in the **water cycle**. The water cycle is the movement of water from Earth's surface to the air and back again.

As water moves through the water cycle, it changes form. When water gains heat, it changes from a liquid to the gas water vapor. Then the water vapor cools and changes back to liquid.

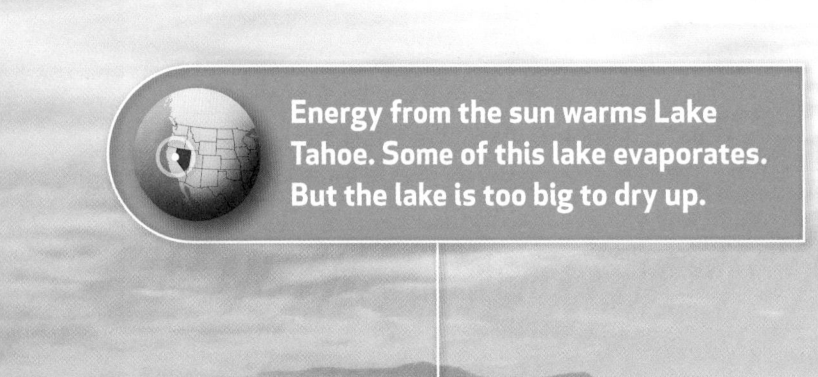

Energy from the sun warms Lake Tahoe. Some of this lake evaporates. But the lake is too big to dry up.

Water vapor is water that has evaporated. Evaporation requires heat. The sun's energy is the source of this heat. During the day the sun shines on oceans, lakes and rivers. The water heats up and some of it evaporates. Energy from the sun causes the water cycle.

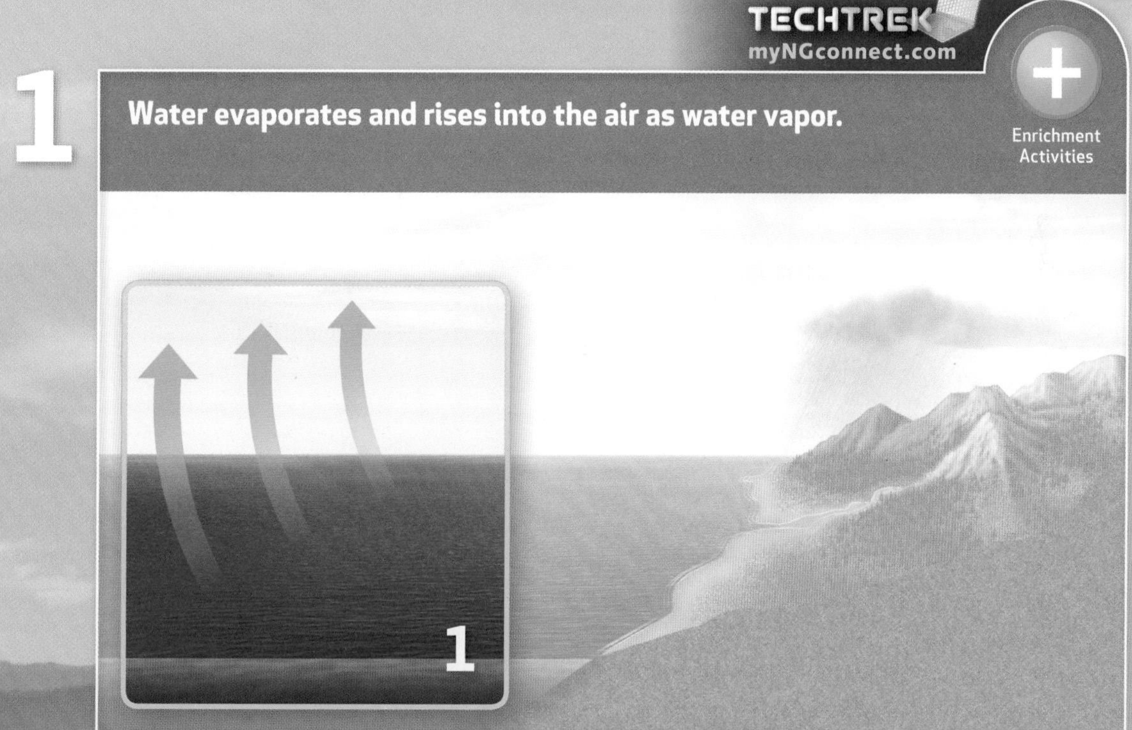

As the water vapor rises, it cools. Condensation turns the water vapor from a gas into tiny drops of water. If the air is cold enough, the water vapor turns into tiny crystals of ice. The ice pieces or water droplets in the air form clouds.

2

Water vapor cools as it rises. Condensation then forms clouds.

Ice crystals or water droplets in clouds can swirl around and grow. Droplets grow by crashing into each other. Crystals grow as more water freezes onto them. When the droplets or crystals get large enough, they fall to the ground as rain or snow. This is called precipitation. Observe different kinds of precipitation in the chart below.

3
Precipitation then falls from clouds as rain, snow, sleet, or hail.
3

TYPES OF **PRECIPITATION**

Water falls to the ground as precipitation. Most of that water just evaporates. Some of it soaks through the surface of Earth and becomes groundwater. Groundwater is water that collects in cracks and soil below the surface. Springs occur where groundwater flows through the surface. Sometimes people dig wells to use groundwater.

Precipitation returns some of the water to lakes, rivers, and oceans.

4

Some runoff flows into low spots to form pools or puddles.

The rest of the water becomes runoff. Runoff is water that flows over Earth's surface in streams and rivers. Rivers and streams flow over the land into other bodies of water such as oceans, lakes, or ponds.

This is the Godafoss waterfall in Iceland. Waterfalls occur wherever rivers flow over cliffs.

Before You Move On

1. What are the main steps of the water cycle?
2. Why is the water cycle important?
3. Analyze What part do clouds play in the water cycle?

OLD WATER

How old is Earth's water? Remember that Earth's water is always moving. It goes from Earth's surface to clouds, and back to Earth again. It can flow in a river one day—and then splash you in the shower the next.

But Earth's water supply does not grow or shrink. Observe the water in the drinking fountain below. George Washington could have sipped this same water in a cup of tea. Dinosaurs could have splashed through the same water millions of years ago. All water on Earth is recycled, but it is never new. Think about this the next time you take a drink of water.

Dinosaurs lived more than 65 million years ago, but they drank the same water you drink today.

Conclusion

Water covers most of Earth's surface. As it moves through the water cycle, it can change to a solid, a liquid, or a gas. Earth's limited supply of water is recycled through the water cycle.

Big Idea Water changes form as it moves through the water cycle.

Vocabulary Review

Match the following terms with the correct definition.

A. salt water
B. fresh water
C. evaporation
D. condensation
E. freezing
F. water cycle

1. The change from a liquid to a gas
2. The movement of water from Earth's surface to the air and back again
3. The change from a liquid to a solid
4. Water that has little or no salt
5. The change from a gas to a liquid
6. Water that has dissolved salts in it

Big Idea Review

1. Recall What are the three forms of water?

2. List What are four types of precipitation?

3. Contrast How are evaporation and condensation different?

4. Classify Classify the bodies of water as salt water or fresh water: glacier, lake, ocean, river

5. Draw Conclusions How would the water cycle be different without the energy of the sun?

6. Generalize Is the water you used this morning the same water dinosaurs may have used millions of years ago? Explain your answer.

my SCIENCE notebook

Write About the Water Cycle

Predict This photo shows clouds and rain. Predict some things that could happen to the rain water once it reaches Earth's surface.

NATIONAL GEOGRAPHIC

CHAPTER 8

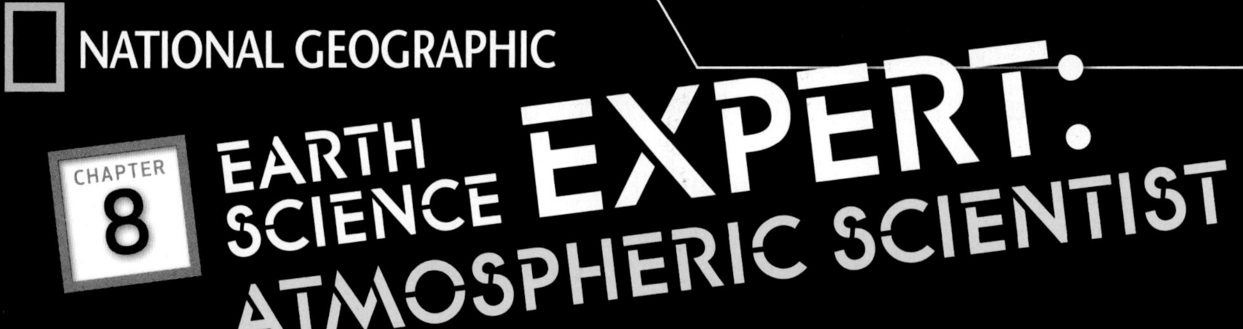

EARTH SCIENCE EXPERT: ATMOSPHERIC SCIENTIST

What's it like to predict earth's future climate? Ask an atmospheric scientist!

Warren Washington loved science when he was a kid in school. But he didn't think he could be a scientist. Then he started to read books about the lives of famous scientists. "I found out they had ordinary lives." Washington says. "This gave me the idea that I could be a scientist, too."

Washington stands at the South Pole in Antarctica.

Washington became an atmospheric scientist. That is a scientist who studies weather and climate. He creates and works with computer models of the atmosphere. A computer model is a huge computer program. It models the gases and winds of the atmosphere. Scientists like Washington use models to learn how the atmosphere works.

Would you like to be an atmospheric scientist? You will need a college degree. You will also take classes in physics and chemistry. It will help if you like math and computers. Like Washington, many atmospheric scientists work for research centers or universities. They must write and speak well to share their work with others.

TECHTREK
myNGconnect.com
Digital Library

Washington hopes to inspire students to choose careers in science.

NATIONAL GEOGRAPHIC

BECOME AN EXPERT

The Mississippi: A River at Work

The Mississippi River is one of the longest rivers in the world. It flows 3780 kilometers (2350 miles) from Minnesota's Lake Itasca to the Gulf of Mexico. Along the way, people fish and play in it. Farmers use its **fresh water** for their crops. Huge barges carry goods up and down the river. Travel down the Mississippi, and see a great river at work!

The Mississippi River starts in Lake Itasca. The river is small and shallow here. You could walk across it.

fresh water

Fresh water is water that has little or no salt in it.

TECHTREK
myNGconnect.com
Student eEdition
Digital Library

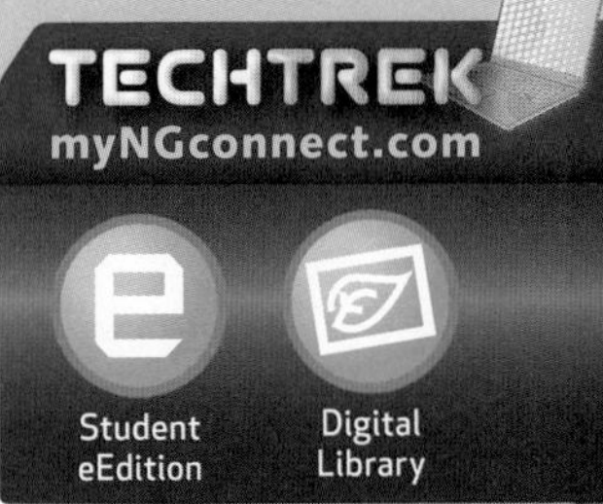

Water for Electricity

Downstream from Lake Itasca, the Mississippi flows through Minneapolis. Flowing water can be used to produce electricity. Here, the river tumbles over a falls. The energy of the falling water is changed into electricity for the city. People in Minneapolis have used the power of the river for almost 200 years.

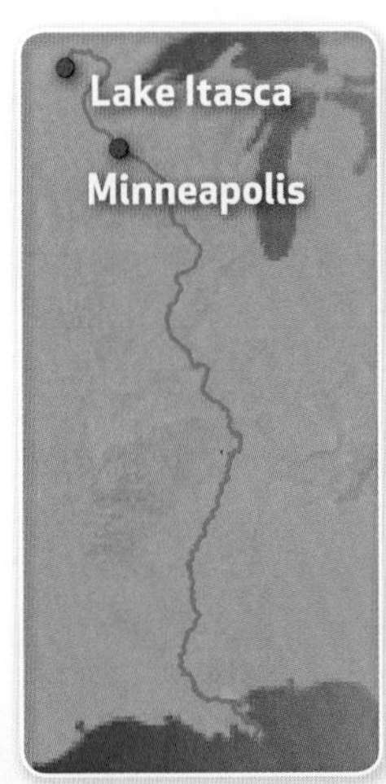

The Mississippi drops almost 13 meters (40 feet) at St. Anthony's Falls in Minneapolis.

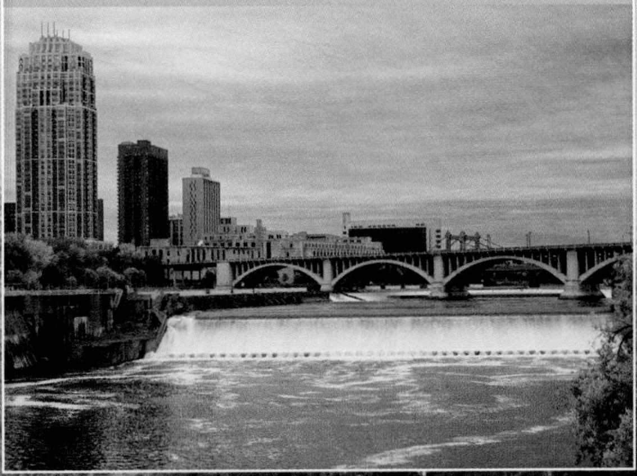

Water for Fun

Millions of people use the Mississippi for fun. They swim and play in it. They enjoy boating and fishing in its slow-moving water. They hike and camp along its banks. On the northern parts of the river, **freezing** temperatures turn the water to ice. The ice can get thick enough for ice skating.

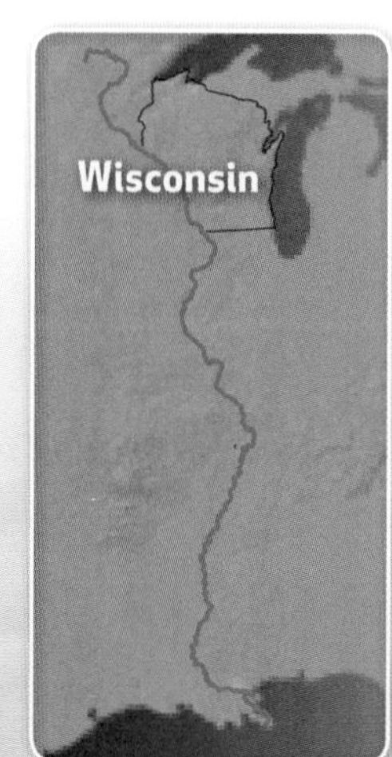

Boating is popular in Wisconsin and other states along the river.

freezing

Freezing is the change from a liquid to a solid.

Water for Drinking

Do you know where your last drink of water came from? More than 50 cities get their drinking water from the Mississippi River. People cannot drink straight from the river. Bacteria in the water can make them sick. So the water is first sent to a treatment plant. The plant makes the water safe to use.

More than 18 million people drink water from the Mississippi River and the smaller rivers that flow into it.

Water for Wildlife

The Mississippi is home to many kinds of fish, birds, and other wildlife. So are the hundreds of smaller rivers flowing into it. The Mississippi collects water from almost half of the surface of the United States. Some water is lost through **evaporation**. But the river grows bigger as more water empties into it. As the river grows, more and more living things depend on it to survive.

These trumpeter swans depend on the river for food and a safe place to land.

evaporation

Evaporation is the change from a liquid to a gas.

Water for Crops

When the **water cycle** produces more **condensation** and rain than usual, the Mississippi can flood. The flood waters make soil richer. Farmers grow crops in this rich, dark soil. They also use the river to water crops during dry times.

Farmers grow crops on the flat lands along the river.

condensation

Condensation is the change from a gas to a liquid.

water cycle

The **water cycle** is the movement of water from Earth's surface to the air and back again.

Water for Shipping

A lot of people use the Mississippi in the same way they use a road—to move things and to get to places. Instead of cars and trucks, they use boats and barges. For example, thousands of boats and barges float to and from New Orleans each year. They carry food and other products from all over the United States.

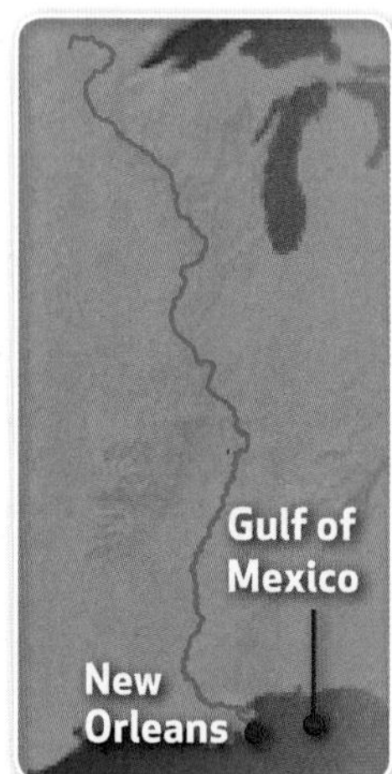

At New Orleans, the river is almost at its end. It is now very wide and about 67 meters (200 feet) deep.

This ship carries goods around the world. Here it unloads its cargo at the port of New Orleans.

Water for the Ocean

The great river ends at the Mississippi River delta. This is where the river's fresh water empties into the **salt water** of the Gulf of Mexico. The river is important for the land and for the ocean. Its delta protects the land behind it from ocean storms. Meanwhile, its fresh water carries food and other materials to ocean life in the Gulf.

TECHTREK myNGconnect.com

Digital Library

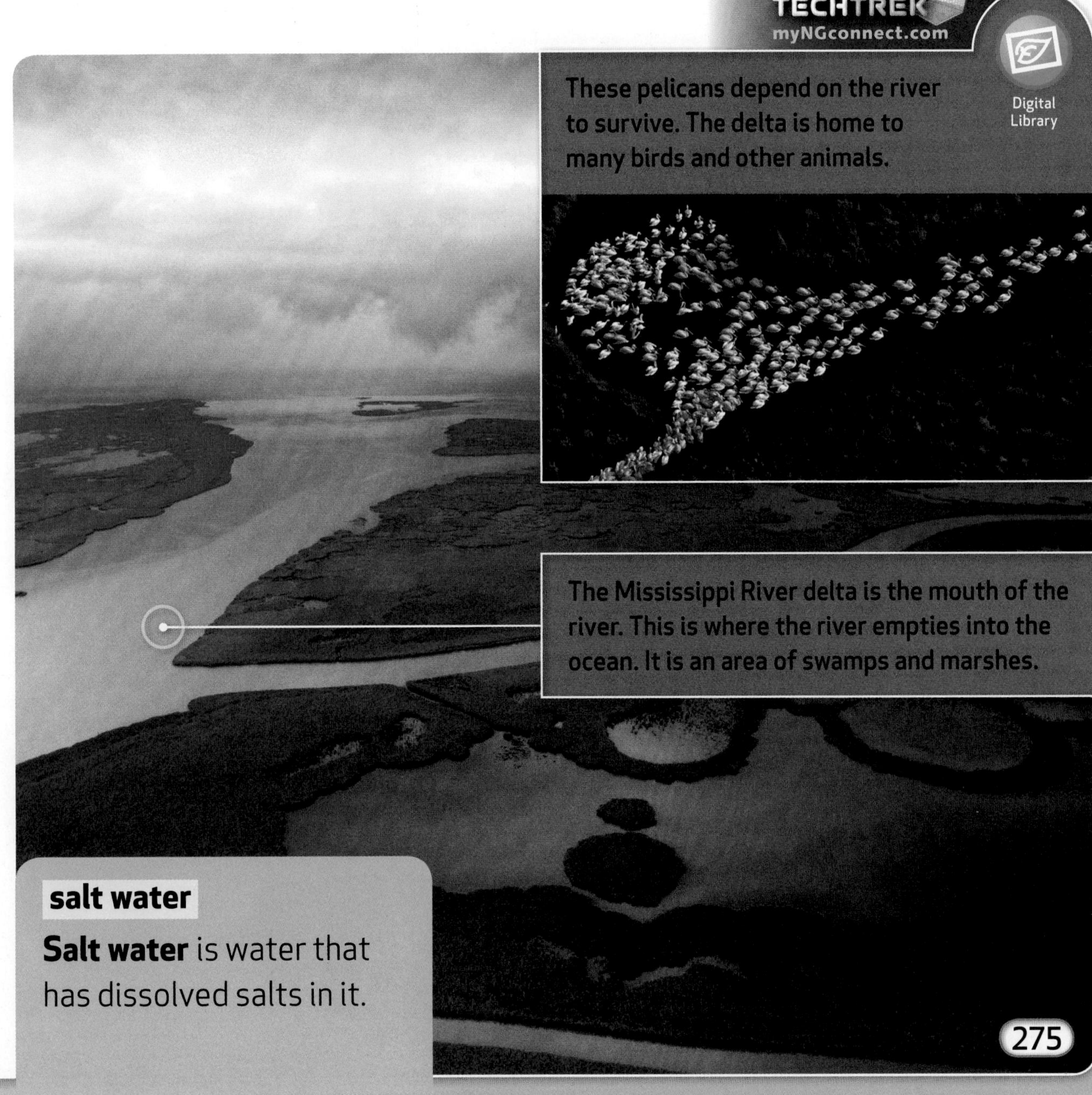

These pelicans depend on the river to survive. The delta is home to many birds and other animals.

The Mississippi River delta is the mouth of the river. This is where the river empties into the ocean. It is an area of swamps and marshes.

salt water

Salt water is water that has dissolved salts in it.

BECOME AN **EXPERT**

NATIONAL GEOGRAPHIC

CHAPTER 8 SHARE AND COMPARE

Turn and Talk How do people use the Mississippi River? Form a complete answer to this question together with a partner.

Read Select two pages in this section. Practice reading the pages. Then read them aloud to a partner. Talk about why the pages are interesting.

my SCIENCE notebook **Write** Write a conclusion that tells the important ideas about the Mississippi River. State what you think is the Big Idea of this section. Share what you wrote with a classmate. Compare your conclusions.

my SCIENCE notebook **Draw** Draw a picture of a summer activity on the Mississippi River. Combine your drawing with those of your classmates to plan a fun weekend trip along the Mississippi.

Glossary

B

brightness (BRĪT-nes)
Brightness is the amount of light that reaches your eye from an object such as a star. (p. 76)

The brightness of a light depends partly on its distance from the viewer.

C

condensation (kon-den-SĀ-shun)
Condensation is the change from a gas to a liquid. (p. 254)

D

deposition (dep-uh-ZISH-uhn)
Deposition is the laying down of rock and soil in a new place. (p. 185)

E

earthquake (URTH-kwāk)
An earthquake is the shaking of the ground caused by the movement of Earth's crust. (p. 210)

energy (EN-ur-jē)
Energy is the ability to do work or cause a change. (p. 44)

erosion (ē-ROH-zhun)
Erosion is the picking up and moving of rocks and soil to a new place. (p. 182)

evaporation (i-vap-uh-RĀ-shun)
Evaporation is the change from a liquid to a gas. (p. 254)

F

fossil fuel (FOS-il FYŪ-el)
A fossil fuel is a source of energy formed from plants or animals that died long ago. (p. 146)

Moving water causes erosion by carrying rocks and soil away.

freezing (FRĒz-ing)
Freezing is the change from liquid to solid. (p. 252)

fresh water (FRESH WAH-ter)
Fresh water is water that has little or no salt. (p. 251)

G

glacier (GLĀ-shur)
A glacier is a huge area of ice that moves slowly over Earth's surface. (p. 184)

grain (GRĀN)
A grain is a piece that makes up rock or soil. (p. 109)

H

humus (HYŪ-mus)
Humus is a part of soil made of bits of decayed plants and animals. (p. 114)

L

landform (LAND-form)
A landform is a natural feature on Earth's surface. (p. 186)

lava (LAH-vah)
Lava is melted rock that comes to Earth's surface through a volcano. (p. 215)

light (LĪT)
Light is a kind of energy you can see. (p. 45)

M

magma (MAG-mah)
Magma is melted rock below Earth's surface. (p. 215)

mineral (MIN-ur-ul)
Minerals are solid nonliving materials found in nature. (p. 108)

The mineral malachite is always green.

N

natural resources
(NACH-ur-al RĒ-sors-es)
Natural resources are living and nonliving things found on Earth that people need. (p. 138)

nonrenewable resources
(non-rē-NŪ-ah-bl RĒ-sors-es)
Nonrenewable resources are those that cannot be replaced quickly enough to keep from running out. (p. 144)

O

orbit (OR-bit)
An orbit is the path Earth or another object takes in space as it revolves. (p. 10)

P

phase (FĀZ)
A phase is a lighted shape of the moon that we see from Earth. (p. 18)

plate (PLĀT)
A plate is a huge piece of Earth that floats on a layer of liquid rock. (p. 211)

property (PROP-ur-tē)
A property is something about an object that you can observe. (p. 76)

R

recycling (rē-SĪ-kling)
Recycling is using the materials in an old object to make a new object. (p. 154)

renewable resources
(rē-NŪ-ah-bl RĒ-sors-es)
Renewable resources are materials that are always being replaced and will not run out. (p. 140)

revolve (ri-VAWLV)
To revolve is to travel around another object in space. (p. 10)

rotate (RŌ-tāt)
To rotate is to spin around. (p. 12)

S

salt water (SAWLT WAH-ter)
Salt water is water that has dissolved salts in it. (p. 251)

satellite (SAT-il-īt)
A satellite is an object that revolves around a larger object. (p. 16)

soil (SOY-il)
Soil is a layer of loose materials on Earth's surface that is made up of bits of rocks, humus, air, and water. (p. 107)

star (STAR)
A star is a glowing ball of hot gases. (p. 74)

sun (SUN)
The sun is the star that is nearest to Earth. (p. 42)

The sun looks bigger and brighter than other stars because it is the nearest star to Earth.

T

telescope (TEL-uh-scōp)
A telescope is a tool that magnifies objects and makes them look closer and bigger. (p. 82)

temperature (TEM-pur-ah-chur)
Temperature is a measure of how hot or cold something is. (p. 49)

transform (trans-FORM)
To transform is to change. (p. 47)

V

volcano (vol-KĀ-nō)
A volcano is an opening in Earth's crust through which lava, gases and ash erupt. (p. 214)

Lava flows out of this volcano.

W

water cycle (WAH-ter SĪ-kel)
The water cycle is the movement of water from Earth's surface to the air and back again. (p. 256)

weathering (WETH-urh-ing)
Weathering is the breaking apart, wearing away, or dissolving of rocks. (p. 178)

Weathering by wind helped to shape this rock.

Index

S

T

Credits

Front Matter

About The Cover (bg) G. Brad Lewis Photography. (t inset) G. Brad Lewis Photography. (b inset) G. Brad Lewis/Science Faction/Corbis. **ii–iii** (bg) Alaska Stock Images/National Geographic Image Collection. **iv–v** Robert Clark/National Geographic Image Collection. **vi–vii** (bg) Jim Richardson/National Geographic Image Collection. **vii** (t) Bill Bachmann/Photo Network/Alamy Images. **viii–ix** (bg) Digital Vision./Getty Images. **ix** (t) Nina Shannon/iStockphoto. **x–1** Alaska Stock Images/National Geographic Image Collection. **2** (t) NASA/JPL–Caltech/T. Pyle (SSC)/Jet Propulsion Laboratory. (c) Don Mason / Blend Images/Getty Images. (b) Gemini Observatory/Ingrid Braul, Southlands Elementary, Vancouver BC. **3** (t) RF Company/Alamy Images. (ct) Neale Cousland/Shutterstock. (c) Lori Adamski Peek/Getty Images. (cb) Frederic J. Brown/AFP/Getty Images. (b) Darren Baker/Alamy Images. **4** (l) NASA/Bill Ingalls. (r) Jay Friedlander/NASA Goddard Space Flight Center.

Chapter 1

5, 6–7 Christian Heinrich/imagebroker/Alamy Images. **8** (t, c) NASA/JPL-Caltech/T. Pyle (SSC)/Jet Propulsion Laboratory. (b) NASA/Goddard Space Flight Center Scientific Visualization Studio, and Digital Globe/QuickBird; Space Imaging/IKONOS; MODIS Rapid Response Team, NASA/GSFC. **9** (b) George Ostertag/age fotostock. (t) Robert F. Balazik/Shutterstock. **10–11** NASA/JPL-Caltech/T. Pyle (SSC)/Jet Propulsion Laboratory. **12** (inset) NASA/Goddard Space Flight Center Scientific Visualization Studio, and Digital Globe/QuickBird; Space Imaging/IKONOS; MODIS Rapid Response Team, NASA/GSFC. **12–13** (bg) Karl Kost/Alamy Images. **13** (b) Jim Zuckerman/Alamy Images. (c) Sailorr/Shutterstock. **14–15** (bg) David Liu/iStockphoto. **15** (1, r) Andrew Northrup. **16** (c) Robert F. Balazik/Shutterstock. **16–17** (bg) Fancy/Alamy Images. **17** (b) Alex Staroseltsev/Shutterstock. **18–19** (bg) George Ostertag/age fotostock. **19** (inset) PhotoDisc/Getty Images. **20–21** (b) Scott Sroka/National Geographic Image Collection. (t) William Radcliffe/Science Faction/Corbis. **22** Judy Waytiuk/Alamy Images. **22–23** (bg) All Canada Photos/Alamy Images. **24** (c) Alex Staroseltsev/Shutterstock. **24–25** (bg) Michael S. Quinton/National Geographic Image Collection. **25** George Ostertag/age fotostock. **26** (t) Rune Thorstein/Flickr/Getty Images. (b) NASA Goddard Space Flight Center. **26–27** (bg) PhotoDisc/Getty Images. **28–29** (bg) Mark Newman/Photo Researchers, Inc. **29** (cl) NASA/Goddard Space Flight Center Scientific Visualization Studio, and Digital Globe/QuickBird; Space Imaging/IKONOS; MODIS Rapid Response Team, NASA/GSFC. **30** (cl) NASA/Goddard Space Flight Center Scientific Visualization Studio, and Digital Globe/QuickBird; Space Imaging/IKONOS; MODIS Rapid Response Team, NASA/GSFC. **30–31** (bg) Christoforos Giatzakis/Alamy Images. **31** (inset) Tony Kurdzuk/Star Ledger/Corbis. **32–33** (bg) SuperStock. **33** (cl) NASA/Goddard Space Flight Center Scientific Visualization Studio, and Digital Globe/QuickBird; Space Imaging/IKONOS; MODIS Rapid Response Team, NASA/GSFC. **34** (cl) NASA/Goddard Space Flight Center Scientific Visualization Studio, and Digital Globe/QuickBird; Space Imaging/IKONOS; MODIS Rapid Response Team, NASA/GSFC. **34–35** (bg) Cusp/SuperStock. **36** Christoforos Giatzakis/Alamy Images.

Chapter 2

37, 38–39 Michael Stubblefield/Alamy Images. **40** (t) Alexandr Tovstenko/iStockphoto. (c) SOHO (ESA & NASA). (b) Thomas Barwick/Lifesize/Getty Images. **41** (t) Bill Bachmann/Photo Network/Alamy Images. (b) Bill Wymar/Alamy Images. **42** (inset) Image courtesy NASA/JPL/Texas A&M/Cornell. **42–43** (bg) Alexandr Tovstenko/iStockphoto. **44–45** SOHO (ESA & NASA). **46–47** (bg) Bill Bachmann/Photo Network/Alamy Images. **47** (l) Dallas and John Heaton/SCPhotos/Alamy Images. (r) Russell Kord/Alamy Images. **48** (t) Thomas Barwick/Lifesize/Getty Images. (b) Bill Wymar/Alamy Images. **48–49** (bg) Thomas Fredriksen/Shutterstock. **49** (l, r) Andrew Northrup. **50–51** (bg) NASA/TRACE/Lockheed Martin Solar & Astrophysics Laboratory. **51** (l) SOHO (ESA & NASA). (r) Solar-B Project/NAOJ. **52** (inset) Christopher Elwell/Shutterstock. **52–53** (bg) tororo reaction/Shutterstock. **53** (tl) Chase Swift/Corbis. (tr) Rafa Irusta/Shutterstock. (bl) Taylor S. Kennedy/National Geographic Image Collection. (br) Todd Arbini Photography/iStockphoto. **54** (l) Bill Barksdale/AGStockUSA/Alamy Images. (r) GSK/Shutterstock. **54–55** (bg) Fedorov Oleksiy/Shutterstock. **55** (l) David Papazian/Corbis. (r) DAJ/White Rock/Getty Images. **56** (l) Alexandr Tovstenko/iStockphoto. (r) Daniel Simon/Westend61/Corbis. **57** (bg) SOHO (ESA & NASA). (inset) Nadiya/Shutterstock. **58** 2009 NREL. All rights reserved. **58–59** (bg) Ellen McKnight/Alamy Images. **60** Iakov Kalinin/Shutterstock. **60–61** (bg) Boris Ryaposov/Shutterstock. **62** Steven Newton/Shutterstock. **62–63** (bg) Karl Weatherly/Corbis. **64** James Davis Photography/Alamy Images. **65** Norbert Rosing/National Geographic Image Collection. **66** Alaska Stock Images/National Geographic Image Collection. **66–67** (bg) NASA Human Space Flight Gallery. **68** Karl Weatherly/Corbis.

Chapter 3

69, 70–71 Allan Morton/Photo Researchers, Inc. **72** (t) Jim Richardson/National Geographic Image Collection. **73** (t) Jason Edwards/National Geographic Image Collection. (b) Stuart O'Sullivan/Getty Images. **74** SIME/eStock Photo. **74–75** (bg) M. Eric Honeycutt/iStockphoto. **75** (l) Richard Nowitz/National Geographic Image Collection. (r) Jim Richardson/National Geographic Image Collection. **76–77** (bg) Jason Edwards/National Geographic Image Collection. **78–79** (bg) traumlichtfabrik/Flickr/Getty Images. **80** Scott Johnson/Starfire Studios. **81** (t) John Chumack/Photo Researchers, Inc. (ct) NASA, ESA, G. Bacon (STScI). (cb) SOHO (ESA & NASA). (b) Andrea Dupree (Harvard-Smithsonian CfA), Ronald Gilliland (STScI), NASA and ESA. **82** (l) Stephen Alvarez/National Geographic Image Collection. **82–83** (bg) Stuart O'Sullivan/Getty Images. **83** (l, r) Andrew Northrup. **84** (inset) Gemini Observatory/Ingrid Braul, Southlands Elementary, Vancouver BC. **84–85** (bg) Gemini Observatory - Northern Operations Center. **85** Neelon Crawford/Polar Fine Arts/Gemini Observatory/AURA. **86** (bg) digipixpro/iStockphoto. **87** (t) Stocktrek Images, Inc./Alamy Images. (b) Long Beach Press-Telegram, Wally Pacholka/AP Images. **88** (l) Russell Croman/Photo Researchers, Inc. (r) Scott Johnson/Starfire Studios. **88–89** (bg) Diego Barucco/Shutterstock. **89** Russell Croman/Photo Researchers, Inc. **90** Space Telescope Science Institute. **91** NASA, ESA, H. Richer (UBC), J. Kalirai (UCSC)/Space Telescope Science Institute. **92–93** (bg) NASA, STScI. **93** NASA, ESA, and The Hubble Heritage Team (STScI/AURA). **94** NASA, ESA and J. Hester (ASU). **96** NASA and The Hubble Heritage Team (AURA/STScI). **97** (bg) The Hubble Heritage Team (AURA/STScI/NASA). (inset) NASA, ESA, P. Challis and R. Kirshner (Harvard-Smithsonian Center for Astrophysics). **98** NASA. **99** NASA, ESA, and The Hubble Heritage Team (STScI/AURA). **100** NASA, ESA and J. Hester (ASU).

Chapter 4

101, 102–103 Paul Van Benschoten/National Geographic Image Collection. **104** (t) Andrew Northrup. (bl) GC Minerals/Alamy Images. (br) RF Company/Alamy Images. **105** (t) PhotoDisc/Getty Images. (b) Kenneth W. Fink/Photo Researchers, Inc. **106–107** nagelestock.com/Alamy Images. **107** (t) Diane Cook and Len Jenshel/National Geographic Image Collection. (ct) Richard Thornton/Shutterstock. (cb) guidocava/Shutterstock. (c, b) Andrew Northrup. **108–109** (bg) Jan Rihak/iStockphoto. **109** PhotoDisc/Getty Images. **110** (tl, tr) GC Minerals/Alamy Images. (bl) RF Company/Alamy Images. (br) efesan/iStockphoto. **111** (l) Peter Arnold, Inc./Alamy Images. (r) Visuals Unlimited/Corbis. **112–113** (bg) Francois Gohier/Photo Researchers, Inc. **112** (l, r) Andrew Northrup. **113** (tl) Gary Ombler/Dorling Kindersley/Getty Images. (tr) Visuals Unlimited/Getty Images. (cl) Scientifica/Visuals Unlimited, Inc. (cr) Sean Curry/iStockphoto. (bl) The Natural History Museum/Alamy. (br) Joyce Photographics/Photo Researchers, Inc. **114** Kenneth W. Fink/Photo Researchers, Inc. **114–115** (bg) David Wall/Alamy Images. **116–117** (bg, insets) Andrew Northrup. **118** (t) David R. Frazier Photolibrary, Inc./Alamy Images. (b) Kenneth Garrett/National Geographic Image Collection. **119** (t) Rick D'Elia/Corbis. (b) Warren Morgan/Corbis. **120** (l) GC Minerals/Alamy Images. (c) Joyce Photographics/Photo Researchers, Inc. (r) Sergey Galushko/Alamy Images. **120–121** (bg) Weeping Willow Photography/Flickr/Getty Images. **121** Astrid & Hanns-Frieder Michler/Photo Researchers, Inc. **122–123** (bg) Chicago Botanic Gardens. **123** (t) Jakub Krechowicz/Shutterstock. (b) Chicago Botanic Gardens. **124–125** Richard Nowitz/National Geographic Image Collection. **126–127** (bg) Jim Richardson/

Chapter 5

133, 134–135 James L. Stanfield/National Geographic Image Collection. **136** (t) iofoto/Shutterstock. (b) Annie Griffiths Belt/ National Geographic Image Collection. **137** (t) Spencer Grant/ PhotoEdit. (c) artproem/Shutterstock. (b) 81A Productions/ Photolibrary. **138–139** (bg) iofoto/Shutterstock. **139** (t) amana images inc./Alamy Images. (ct) BanksPhotos/iStockphoto. (c) Dex Image/Alamy Images. (cb) Steve Cole/iStockphoto. (b) George F. Mobley/National Geographic Image Collection. **140** (inset) Annie Griffiths Belt/National Geographic Image Collection. **140–141** (bg) Annie Griffiths Belt/National Geographic Image Collection. **142–143** (bg) Ed Kashi/National Geographic Image Collection. **143** Lori Adamski Peek/Getty Images. **144–145** (bg) Spencer Grant/PhotoEdit. **145** (inset) Vladimir Kolobov/iStockphoto. **146** (inset) James L. Stanfield/National Geographic Image Collection. **146–147** (bg) Bounce/UpperCut Images/Getty Images. **148–149** (bg) Digital Vision/Getty Images. **149** (inset) London Aerial Photo Library/Alamy Images. **150–151** (bg) Corbis/Jupiterimages. **151** Photobank/Shutterstock. **152** (inset) Melissa Farlow/National Geographic Image Collection. **152–153** (bg) Melissa Farlow/ National Geographic Image Collection. **154–155** (bg) Phil Degginger/Alamy Images. **154** (l) LyaC/iStockphoto. **155** (tl) LyaC/ iStockphoto. (tr) artproem/Shutterstock. (bl) 81A Productions/ Photolibrary. (br) PixAchi/Shutterstock. **156–157** Corbis. **158–159** (bg) Robert Clark/National Geographic Image Collection. **159** (inset) Markley Boyer/National Geographic Image Collection. **160** (l) Ed Kashi/National Geographic Image Collection. (r) Corbis. **160–167** (bg) Martin Mistretta/The Image Bank/Getty Images. **161** Pete Ryan/National Geographic Image Collection. **162** The Clean Air Campaign. **163** (bg) Corbis. (tl, tr) Noah Strycker/Shutterstock. **164** NASA/Goddard Space Flight Center Scientific Visualization Studio. **164–165** (b) Jim West/Alamy Images. **165** (t) Inmagine/ Alamy Images. **166** (t) Corbis RF/Alamy Images. (bl) Steven Weinberg/Riser/Getty Images. (bc) Claudius/Corbis. (br) Sami Sarkis/PhotoDisc/Getty Images. **167** Petr Nad/Shutterstock. **168** (bg) Digital Vision/Getty Images. (inset) SasPartout/ Shutterstock. **169** (tl) PhotoDisc/Getty Images. (tc) imagebroker/ Alamy Images. (tr) AVAVA/Shutterstock. (b) Phil Schermeister/ National Geographic Image Collection. **170–171** (bg) Lee O'Dell/ Shutterstock. **171** (l) Digital Vision/Getty Images. (c) Oralleff/ Shutterstock. (r) Kraig Scarbinsky/Digital Vision/Getty Images. **172** Digital Vision/Getty Images.

Chapter 6

173, 174–175 Neale Cousland/Shutterstock. **176** (c) Ira Block/ National Geographic Image Collection. **177** (t) Rob Watkins/Alamy Images. (b) Heather Perry/National Geographic Image Collection. **178–179** Melissa Farlow/National Geographic Image Collection. **179** geogphotos/Alamy Images. **180** Digital Stock/Corbis. **182** Larry Miller/Photo Researchers, Inc. **183** (t) Ira Block/National Geographic Image Collection. (bl, br) Andrew Northrup. **184** Gerald & Buff Corsi/Visuals Unlimited. **184–185** (bg) Rob Watkins/ Alamy Images. **186–187** Glen Allison/PhotoDisc/Getty Images. **187** (t) Heather Perry/National Geographic Image Collection. (tc) Glen Allison/PhotoDisc/Getty Images. (bc) Jame P. Blair/National Geographic Image Collection. (b) Tom Bean/Corbis. **188–189** (bg) Raymond Gehman/National Geographic Image Collection. **189** (t) Robert C Nunnington/Gallo Images/Getty Images. (b) Michael Nichols/National Geographic Image Collection. **190** (t) DAJ/Getty Images. (b) Kang Khoon Seang/Shutterstock. **191** Jim Richardson/ National Geographic Image Collection. **192** (r) Larry Miller/Photo Researchers, Inc. **192–193** (bg) Ralph Lee Hopkins/National Geographic Image Collection. **193** Konrad Wothe/Minden Pictures/ National Geographic Image Collection. **194** (bg) offiwent.com/ Alamy Images. (inset) Phil Stoffer/United States Geological Survey staff /life.nbii.gov. **195** Phil Stoffer/United States Geological Survey staff /life.nbii.gov. **196** Melville B. Grosvenor/National Geographic Image Collection. **196–197** (bg) Tim Laman/National Geographic Image Collection. **197** (t) Scott Hailstone/iStockphoto. **198** Barry Bishop/National Geographic Image Collection. **200–201** Oliver Gerhard/Imagebroker RF/age fotostock. **201** Stephen Alvarez/National Geographic Image Collection. **202–203** (bg) David S. Boyer and Arlan R. Wiker/National Geographic Image Collection. **203** Marc Muench/Alamy Images. **204** Oliver Gerhard/Imagebroker RF/age fotostock.

Chapter 7

205, 206–207 Chip Somodevilla/Getty Images. **208** (t) Frederic J. Brown/AFP/Getty Images. (b) Kevin Schafer/Alamy Images. **209** (c) Picture Press/Alamy Images. (b) Bullit Marquez/AP Images. **210–211** Kevin Schafer/Alamy Images. **210** (l, r) Andrew Northrup. **212–213** (bg) Frederic J. Brown/AFP/Getty Images. **213** Karen Kasmauski/Corbis. **214** Bullit Marquez/AP Images. **214–215** Picture Press/Alamy Images. **216** Vincent J. Musi/National Geographic Image Collection. **216–217** (bg) Kevin West/Getty Images. **217** Vincent J. Musi/National Geographic Image Collection. **218** G. R. 'Dick' Roberts/ NSIL/Visuals Unlimited/Getty Images. **219** Ed Harp, USGS. **220** (l) DBURKE/Alamy Images. (r) Mike Theiss/National Geographic Image Collection. **220–221** (bg) Creatas/Jupiterimages. **221** Losevsky Pavel/Shutterstock. **222** (bg) Annie Griffiths Belt/National Geographic Image Collection. (inset) The Stocktrek Corp/Brand X Pictures/Jupiterimages. **223** Carsten Peter/National Geographic Image Collection. **224–225** Marvin Nauman/FEMA. **225** Mark Thiessen/National Geographic Image Collection. **226** (t) rodho/ Shutterstock. (b) DigitalGlobe via Getty Images. (bg) James P. Blair/ National Geographic Image Collection. **227, 228, 229** James P. Blair/National Geographic Image Collection. **230** (l) Bullit Marquez/ AP Images. (c1) David Guttenfelder, File/AP Images. (cr) Ed Harp, USGS. (r) Marvin Nauman/FEMA. **230–231** (bg) Digital Vision./Getty Images. **231** imagebroker/Alamy. **232–233** Stanford University. **234** Robert Clark/National Geographic Image Collection. **234–235, 236** (bg) O. Louis Mazzatenta/National Geographic Image Collection. **237** (t, b) Richard Nowitz/National Geographic Image Collection. **238** Louis S. Glanzman/National Geographic Image Collection. **238–239** (bg) O. Louis Mazzatenta/National Geographic Image Collection. **240** (t) Richard Nowitz/National Geographic Image Collection. (b) Roger Ressmeyer/Corbis. **241** (t) James L. Stanfield/ National Geographic Image Collection. (b) Jonathan Blair/National Geographic Image Collection. **242–243** Sailorr/Shutterstock. **244** Robert Clark/National Geographic Image Collection.

Chapter 8

245, 246–247 Karl Weatherly/Digital Vision/Getty Images. **248** (t) Stocktrek Images, Inc./Alamy Images. (c) Radius Images/ Alamy Images. (b) Kord.com/age fotostock. **250–251** (bg) Stocktrek Images, Inc./Alamy Images. **251** (t) W. Robert Moore/National Geographic Image Collection. (c) J. Baylor Roberts/National Geographic Image Collection. (b) VisionsofAmerica/Joe Sohm/Digital Vision/ Getty Images. **252** Kord.com/age fotostock. **253** Radius Images/ Alamy Images. **254–255** (bg) Robert & Jean Pollock/Visuals Unlimited. **255** (l, r) Andrew Northrup. **256–257** Adilson Simoes/Flickr/ Getty Images. **258–259** (bg) Royalty-Free/Corbis. **259** (l) Gordon Wiltsie/National Geographic Image Collection. (cl) Tetra Images/ Getty Images. (cr) Andy Hallam/Alamy Images. (r) Harnett/Hanzon/ Photodisc/Getty Images. **260** (inset) Tomasz Parys/Shutterstock. **260–261** (bg) Darren Baker/Alamy Images. **262–263** (bg) Raúl Martín. **263** Nina Shannon/iStockphoto. **264–265** (bg) Per Makitalo/ Getty Images. **265** Raul Touzon/National Geographic Image Collection. **266** National Center For Atmospheric Research. **266–267** (bg) Brand X Pictures/Photolibrary. **267** National Center For Atmospheric Research. **268–269** (bg) Jason Lindsey/Alamy Images. **269** (inset) Reimar Gaertner/Alamy Images. **270–271** (bg) Brent Frazee/Kansas City Star/MCT/NewsCom. **271** AGB Photo/ Alamy Images. **272** Tammy Wolfe/iStockphoto. **273** Grant Heilman/ Grant Heilman Photography/Alamy Images. **274** (bg) Tyrone Turner/National Geographic Image Collection. (inset) Ed Metz/ Shutterstock. **275** (bg) Cameron Davidson/Photographer's Choice RF/Getty Images. (inset) Annie Griffiths Belt/Corbis. **276** Tyrone Turner/National Geographic Image Collection.

Back Matter

EM1 (t) Jason Edwards/National Geographic Image Collection. (b) Larry Miller/Photo Researchers, Inc. **EM2** RF Company/Alamy Images. **EM3** Alexandr Tovstenko/iStockphoto. **EM4** (t) Bullit Marquez/AP Images. (b) Melissa Farlow/National Geographic Image Collection. **EM7** Richard Nowitz/National Geographic Image Collection. **EM8** Annie Griffiths Belt/National Geographic Image Collection. **EM10–EM11** nagelestock.com/Alamy Images. **EM15** Digital Vision/Getty Images. **EM16** NASA, STScI. **Back Cover** (bg) G. Brad Lewis Photography. (tl) NASA/Bill Ingalls. (tr) NASA Goddard Space Flight Center. (c) SOHO (ESA & NASA). (bl) Jay Friedlander/ NASA Goddard Space Flight Center. (br) NASA/TRACE/Lockheed Martin Solar & Astrophysics Laboratory.